Forest and Plant Bioproducts Division 2016

AA002654

Core Programming Area at the 2016 AIChE Annual Meeting

San Francisco, California, USA
13-18 November 2016

ISBN: 978-1-5108-3437-8

Printed from e-media with permission by:

Curran Associates, Inc.
57 Morehouse Lane
Red Hook, NY 12571

Some format issues inherent in the e-media version may also appear in this print version.

Copyright© (2016) by AIChE
All rights reserved.

Printed by Curran Associates, Inc. (2017)

For permission requests, please contact AIChE
at the address below.

AIChE
120 Wall Street, FL 23
New York, NY 10005-4020

Phone: (800) 242-4363
Fax: (203) 775-5177

www.aiche.org

Additional copies of this publication are available from:

Curran Associates, Inc.
57 Morehouse Lane
Red Hook, NY 12571 USA
Phone: 845-758-0400
Fax: 845-758-2633
Email: curran@proceedings.com
Web: www.proceedings.com

TABLE OF CONTENTS

463912 Effect of Date Pit Based Additive on the Thermal Stability of Water-Based Drilling Fluid: Experimental Studies and Mathematical Modeling

Abstract:

Successful completion of oil and gas wells depends to a large extent on the properties of the drilling fluid used during the operation. The ability of drilling fluids to perform their fundamental functions is measured by certain performance properties including fluid loss prevention, stability under various temperature and pressure operating conditions, and stability against corrosion. Varieties of additives (such as polymers) are usually added to drilling fluid formulation in order to fulfill these property requirements. However, the poor thermal stability of such additives makes it difficult for them to be used in high temperature and high pressure drilling operations. Naturally occurring composites (such as date pit) are a new set of materials that are currently generating interest in varieties of applications. Date pit is a green, readily available and cheap material with superior thermal stability as compared to the polymers. Therefore, the objective of this work is to investigate the effect of date pit particles on the thermal stability of water â€" based drilling fluid formulation.

Infusion method was used to prepare date pit particles/poly (2-acrylamido-2-methyl-1-propanesulfonic acid) (PAMS) composite. Commercial Bentonite (Alumino Silicate Hydrate Montmorillonite) was used as the base material for preparing the water based drilling formulation. The formulation was prepared by adding 22.5g of Bentonite and the composite additive in 400ml of pure water. The percentage weight of the composite additive that was used ranges from 0.5 to 10 wt percent. Thermal stability of the formulations with and without date pit particles was evaluated using thermogravimetric analysis within a temperature range of 30 and 600°C under N_2 environment. Particle size analysis was done using laser light diffraction. FTIR study was also performed to evaluate the interaction between the additives and the bentonite.

In our previous studies, it was observed that the addition of 10wt% composite increased the rheological properties (PV, YP and YVR) by 80, 33 and 35% respectively, decreased the filter cake thickness by 35% and reduced the amount of fluid loss by 44%. The results from the present study revealed that the thermal degradation temperature of pure bentonite, date pit and PAMPS are 455, 250 and 160°C, respectively. Formulations with 10wt% date pit additive can be used up to a temperature of 300°C with only a 8% drop in the thermal stability while the thermal stability of the formulation with only PAMPS (without date pit) additive decreased by 53% at the same temperature as compared to the formulation without additive. Therefore, the prepared composite demonstrated some potential for use as a drilling fluid additive in high temperature environment. This observation is an eye opener into the suitability of date pit based drilling fluid formulation for high temperature and high pressure drilling fluid operations. The research outcome is expected to provide useful information regarding the use of readily available cheap waste materials for oil and gas exploration.

Copyright © American Institute of Chemical Engineers. All rights reserved.

475008 Biocarbon Applications- Current Status and Future Directions

Abstract:

Biobased composites for lighter auto-part applications are the wave of the transportation industries. The corporate average fuel economy (CAFE) standard requires 54.5 mpg fleet average by the year 2025. Carbonaneous and lighter biocomposites are being researched and the resulting new innovations show their potential applications in injection moulded parts for car uses. Around 20 wt.% weight saving is achieved for such newly engineered biocomposites as contrast to talc filled counterparts. Carbon black, a petro-based colorant and filler is being used predominantly in plastics and rubber-based industries that accounts to around $ 20 billion market value globally. A more eco-friendly alternative, named as bio-carbon (from pyrolyzed biomass) is attracting recent attention as a filler/reinforcement in automotive and other plastic/rubber-based industries. Although a low content of ~2-4% of carbon black is used as colorant in many plastic-based products, bio-carbon with appropriate modification can be incorporated comparatively at a considerable higher percent (even 30 to 40 wt. %) in specific plastic based composite materials. Such biobased composites show promising potential as possible substitute for talc/short glass fibre reinforced polymer composites.

Acknowledgements: This research is financially supported by (1) the Ontario Ministry of Agriculture, Food, and Rural Affairs (OMAFRA)-University of Guelph Research Program(Project # 030055, 200399); (2) the Natural Sciences and Engineering Research Council (NSERC), Canada, Discovery Grants (Project # 401111); (3) Agriculture and Agri-Food Canada and Competitive Green Technologies (Project #051910, 052881) and (4) the Ontario Research Fund, Research Excellence Program; Round-7 (ORF-RE07) from the Ontario Ministry of Research and Innovation (MRI), currently known as the Ontario Ministry of Research, Innovation and Science (MRIS) (Project # 052644 and # 052665).

Copyright © American Institute of Chemical Engineers. All rights reserved.

467117 Development of a Pilot-Scale Phytate Extraction System from Ethanol Coproducts

Abstract:

Introduction

Commercial production of fuel ethanol involves breaking down the starch present in corn to simple sugars, followed by the bioconversion of sugar molecules to ethanol catalyzed by yeast. The downstream processing involves the recovery of ethanol by distillation and the residual coproducts are further processed to serve as animal feed. In the dry-grind process, which accounts for over 70% of the total plants in the United States, the residual components are separated after the fermentation and distillation.

When coproducts are fed as an energy source to animals (above 15 to 20% of diet), protein and phosphorus are overfed. The environmental impact of feeding protein and phosphorus-rich materials to animals is related to one of the most pervasive forms of pollution from modern agriculture, as animal manure is applied to the soil. Phytate is the primary storage form of phosphorus and inositol in plants. The bioavailability of phosphorus bound to phytate is low for non-ruminants, and thus the phytate-derived phosphate ingested by these species (e.g. poultry and swine) is largely excreted, resulting in both nutritional deficiencies and environmental pollution. On the other hand, phytate is a widely applied valuable chemical as human nutrition, pharmaceutics, cosmetics, chelating chemical, and the raw material to produce inositol (Vitamin B8). It is primarily produced from rice bran in Asia and imported to the US

This creates a great opportunity for the US ethanol industry because extracting phytate as a new product from corn ethanol coproducts, knowing that corn germ has up to 5% phytate in its composition, can create additional revenue while increasing the feeding value of coproducts and decrease the phosphorus excretion in the animal manure.

Methods

Samples from four companies in the Midwest (Iowa, Illinois, and Minnesota) were obtained. Samples are stored at -20 °C prior to analysis and use.

Total-P and Phosphate-P are determined using P-ash method, in which digested samples are analyzed with a colorimetric assay of P in the digests. Samples are first dried at 105 °C until constant dry weight, and subsequently, they will be ashed at 550 °C. Concentrated HNO_3 (7 mol L^{-1}) will be used to dissolve the inorganic ash residue. Spectrophotometric assays are then performed with the molybdenum blue method. An enzymatic method derived from De Boland, Garner, & O'Dell (1975) is used for determination of phytate phosphorus.

Five macroporous resins (IRA series) and the fine resin (AG-1 series) were evaluated for the phytate adsorption capacity, high selectivity, high purity after desorption and low maintenance requirements. All of the resins selected for this study are alkali-type and have strong phytate adsorption capacity. The selection of the resin is one of the most important factors to be considered in the extraction process because all of the extraction process designs are based upon the resin selection.

The adsorption isotherms predict the maximum adsorption capacity of each resin. They are expressed in amount of substrate adsorbed in equilibrium onto the resin (qe) vs. the residual substrate in solution in equilibrium (Ce). Freundlich, Langmuir and Dubinin-Radushkevich adsorption models were used to estimate isotherm parameters.

Results and discussion

WS, containing undissolved corn fragments and residual yeast cells, present a moisture content of 87.94%. The TS, after the solid-liquid separation from WS, has the moisture content of 95.41%. The CDS, concentrated from TS through series of evaporators in order to reduce its moisture content, achieves the moisture content of 70.48%. On regards to the liquid fraction, mainly due to the increase in suspended solids, viscosity and osmotic pressure, the remaining water in CDS was difficult to remove by evaporation (Belyea et al., 2006), achieving moisture content levels close to 70%. The CDS is then mixed with the wet distillers grains to produce WDGS (M.C. 49.75%), which is then further dried to produce DDGS, at the moisture content of 14.92%. Usually, limited by cost in transportation for marketing, DDGS has a low moisture content ranging from 10% to 15%, which also prevents it to become moldy and unusable. Evaporation and drying are one of the most energy intensive steps in the corn ethanol processing.

WS has a total phosphorus concentration of 12.02 mg/g, of which 40% is found as phytate-phosphorus, at a concentration of 4.81 mg/g. After the first solid-liquid separation, the concentrations of total and phytate phosphorus (23.57 mg/g and 10.11 mg/g, respectively) in the TS are over doubled. The water removal step on TS to produce CDS maintains similar levels for total phosphorus (23.77 mg/g) and a decrease on phytate phosphorus (9.41 mg/g). WDGS and DDGS showed similar concentrations of total phosphorus (10.11 mg/g and 10.59 mg/g, respectively), whereas DDGS had higher levels of phytate than WDGS (4.50 mg/g and 3.27 mg/g, respectively). Comparing the results reported at literature (Alkan-Ozkaynak et al., 2010), similar values were found, especially for fractions of TS (19.4 mg/g), CDS (18.8 mg/g), and WS (11.0 mg/g) in means of total phosphorus.

Phytate is found most concentrated in the small-size range of particles in TS, indicating that it is possible to solubilize or aggregate these small particles. Selecting the appropriate size range of particles is crucial in order to process only the most concentrated fraction, which will allow the most efficient adsorption operation. The fractions of 0.10 mm and below account for about 90% of the total phytate found in TS. Screening experiments were conducted to evaluate the pressure drop required to process TS and 0.10 mm-filtered TS. Our results showed that having a filtration step prior to the adsorption process may be energetically feasible, since this step decreases the required pressured drop of the adsorption process by 55%.

Thin stillage is a phytate-rich ethanol co-product and is the liquid fraction that is being evaluated as the raw material in the phytate extraction. It also has significant presence of phosphate and other types of phosphorus-based groups. Reactive-phosphorus (i.e. free orthophosphate) was also adsorbed onto the resin. The characteristics of the most effective resin are those with the capability to adsorb most of the phytate and the least amount of phosphate, when determining resin selectivity. IRA-400 (0.128 mg phytate/mg resin) and IRA-900 (0.126 mg phytate/mg resin) resins adsorbed the most phytate per mass of resin, but IRA-900 (0.733 phytate phosphorus / total phosphorus) showed more selectivity compared to the IRA-400 (0.661 phytate phosphorus / total phosphorus) resin from this screening experiment.

In order to reduce the operating cost of the adsorption process and to achieve technical feasibility, the optimal resin should have an extended lifespan. A study was conducted to evaluate the regeneration capacity of the 5 macroporous (IRA series) resins using NaCl (2 M) as desorbent and NaOH (150 g/L at a loading rate of 4 %) as a regenerating agent for 4 cycles of operation. The equilibrium concentrations (q_e) of phytate-P (mg/g resin) in all the four cycles for all the resins tested, were within 2.5 standard deviations from the mean, which is an acceptable deviation for this study. The adsorption- desorption capacity and selectivity were similar for all 4 cycles.

References

Alkan-Ozkaynak, A., Karthikeyan, K., & Roa-Espinosa, A. (2010). Reducing phosphorus concentration in animal feed coproducts from the corn distilling industry. *Transactions of the ASABE, 53*(4), 1287-1294.

Belyea, R., Rausch, K., & Tumbleson, M. (2004). Composition of corn and distillers dried grains with solubles from dry grind ethanol processing. *Bioresource Technology, 94*(3), 293-298.

De Boland, A. R., Garner, G. B., & O'Dell, B. L. (1975). Identification and properties of phytate in cereal grains and oilseed products. *Journal of Agricultural and Food Chemistry, 23*(6), 1186-1189.

Copyright © American Institute of Chemical Engineers. All rights reserved.

474771 New Biocomposite Materials from Poly(lactic acid) Blended with Poly(ethylene glycol) and Biobased Carbon

Abstract:

Biobased and biodegradable polymeric materials continue to be a fundamental step towards achieving an environmentally-friendly society. Therefore, poly(lactic acid) (PLA) being the most common biopolymer in the marketplace today is useful as the base component in the formation of future compostable consumer products. The current downsides of PLA being its toughness and low thermal stability are the key reason for its limited applications and a major focus for research and optimization.

In this study, poly(lactic acid) based composites prepared through melt blending with poly(ethylene glycol) (PEG) and biobased carbon were tested for their thermal and mechanical properties. The addition of poly(ethylene glycol) acted as a plasticizer, via greater chain mobility and as a crystallization promoter, by the reduction in energy necessary for chain folding during spherulite formation. The biobased carbon was beneficial in establishing nucleation sites for the poly(lactic acid) upon cooling, increasing the crystallization relative to the neat material. The poly(D-lactic acid) (PDLA) was essential in the development of the PLA stereocomplex between the L- and D- enantiomers. The crystallinity of the polymeric composites was elevated when PEG was included in the system. The heat deflection temperature of the poly(lactic acid) with PEG and PDLA was hindered by the low glass transition temperature of the samples. The mechanical properties of the composites samples are still under characterization and will be incorporated into the results prior to event. It is expected that the tensile strength and Youngâ€™s modulus will be lowered with the inclusion of poly(ethylene glycol), while the elongation at break will be increased. Further investigation of the nothched Izod impact strength will be conducted and optical microscopy for visual representations of the crystallization of the composites. Also, scanning electron microscopy (SEM) will be used to determine the morphology and dispersion of the components throughout the composite in order to correlate all findings.

Acknowledgment: *This research is financially supported by (1) the Ontario Ministry of Agriculture, Food, and Rural Affairs (OMAFRA)-New Directions Research Program (Project # 049528) and (2) the Ontario Ministry of Economic Development and Innovation (MEDI), Canada, Ontario Research Fund, Research Excellence Round 4 program (ORF-RE04) (Project # 050231 and 050289).*

Copyright © American Institute of Chemical Engineers. All rights reserved.

474990 Effect of Butyl Glycidyl Ether Model Co-Monomer on Poly (glycerol succinate) Network and Dynamics for the Design of Multifunctional Hyperbranched Polymers

Abstract:

An original approach has been developed to easily design functional materials from poly (glycerol succinate) (PGS). This strategy is related to the introduction of an epoxidized functional agent during the polyesterification between the glycerol and succinic acid. The impact of this epoxide group on the polymerization chemistry and its hyperbranched network has been modeled by using the butyl glycidyl ether (BGE) as co-monomer. The potential reactions have been confronted with the units topology revealed by 2D NMR correlations. The regioselectivity and the stoichiometry of the system were modified *in situ* by the kinetic control of parallel reactions. As a result, we observed a delay of the gelation and an increase of the polyesterification conversion. The resulting hyperbranched polymers (HBPs) obtained just after gelation present a temperature of glass transition (T_g) of -3.9 °C for PGS and -16.1 °C for Poly(glycerol-co-butyl glycidyl ether) succinate (PGS-co-BGE). This difference was related to the dynamic motion of the BGE butyl tails which play the role of spacer between the polymer chains during the relaxation process. The dependency of the effective activation energy ($E_{\hat{I}\pm}$) through the T_g was calculated by using the advanced isoconversional method. The variation of $E_{\hat{I}\pm}$ was associated to a competition between the cooperative chains motions and the sequestration effect of the hydrogen bonds. This behavior can be potentially generalizable to the plastic glass containing a critical amount of secondary interactions.

Acknowledgment: *This research is financially supportedÂ by (1) the Ontario Ministry of Agriculture, Food, and Rural Affairs (OMAFRA)-New Directions Research Program (Project # 050155); and (2) OMAFRA-University of Guelph Bioeconomy-Industrial Uses Theme (Project # 200283).*

Copyright © American Institute of Chemical Engineers. All rights reserved.

475007 Value Added Biodegradable Materials from Industrial Wastes

Abstract:

Biodegradable plastics and composites can be a solution to the plastic waste pollution issue. With regards to practical biodegradability, the term compostability is a standardized way to identify plastics and composites that undergo biodegradation more than 90 % of their weight in a specific time frame; six month. At the Bioproducts Discovery and Development Centre (BDDC), University of Guelph, research is going on to develop compostable composites formulations from biodegradable plastics such as PLA, PBS, etc., and biomass such as agricultural residues, perennial grasses, and industrial co-products/by-products. This approach has two advantages. First, it can create new outlets for the under-valued biomass and bring new revenue to the farmer. Second, it can result in faster biodegradation of the composite in comparison with the neat plastic, as previous studies at the BDDC confirm that. With the compostability in mind, the mechanical and rheological performance of the composite material is an important factor to consider. Green surface treatments of the biomass and the use of processing aids and compatibilizers are effective ways to enhance or keep balanced the properties of the composite material. To date, there is no commercial compatibilizer based on biodegradable plastics. At the BDDC, maleated bioplastic compatibilizers were produced customized towards the bioplastic matrix and successfully utilized for improving the mechanical properties of the compostable composite materials. On the other hand, hybridization of the fillers of different types of biomass was another approach pursued in this study to benefit from different aspects of different fillers at the same time. The amount of compatibilizer and different fillers was optimized to achieve a balance in mechanical and physical properties. The produced compostable composites were successfully certified as 100 % compostable and used in food packaging applications.

Acknowledgements: *This research is financially supported by (1) the Ontario Ministry of Agriculture, Food, and Rural Affairs (OMAFRA)- University of Guelph Bioeconomy Industrial uses Research Program (Project # 200245, 200245); (2) Agriculture and Agri-Food Canada (AAFC) and Competitive Green Technologies (CGTech) through AgriInnovation (Program Project # 051909, 051910, 052880); and (3) and the Ministry of Economic Development and*

Innovation (MEDI), Ontario Research Fund - Research Excellence Round - 4 (Program Project # 050231 and 050289).

Copyright © American Institute of Chemical Engineers. All rights reserved.

456461 Organosolv and Kraft Lignin: Fractionation and Conversion to Melt Spun and Electrospun Carbon Fibers

Abstract:

The global market for carbon fiber is growing rapidly as potential uses in energy and energy savings currently are expanding. The low cost, potential abundance, and highly aromatic carbon content of lignin, compared to polyacrylonitrile (PAN) which is the main precursor for making carbon fiber, make it an attractive precursor for making low-cost carbon fiber.

Melt spinning and electrospinning are two common methods for making fibers with different characteristic and applications. While melt spun fibers can be used in structural and composite applications, electrospun fibers have potential applications in energy storage and filtration. However, these spinning methods required lignins with different characteristics. For example, electrospinning requires high molecular weight and soluble lignin, while melt spinning requires low molecular weight and fusible lignin. Technical lignins, such as kraft and organosolv, are heterogeneous and contains different molecular weight fractions and functionality. Ultrafiltration and sequential solvent extraction are methods which have been employed for separation of different fractions from lignin.

We performed a sequential solvent extraction, using methanol and methanol/methylene chloride mixture to fractionate softwood kraft lignin and organosolv switchgrass lignin. The solvent extraction resulted in obtaining three lignin fractions, according to their solubility. Lignin samples were evaluated by thermal analysis (DSC and TGA), different spectroscopy method (FTIR and NMR), elemental and compositional analysis. Finally, the lignin fractions were used for making melt spun and electrospun carbon fibers. Softwood kraft lignin had the highest ash content and was not fusible. Although switchgrass lignin was fusible, continuous spinning was not possible. Both switchgrass and softwood kraft lignins were not completely soluble in electrospinning solvent (DMF/methanol). After solvent extraction, the methanol soluble fraction had low T_g and thermal decomposition temperature, while the methanol/methylene chloride soluble fraction had high T_g and thermal decomposition temperature (117 vs 181°C for softwodd kraft lignin and 107 vs 161°C for switchgrass lignin) which could indicate low and high molecular weight of these fractions. Methanol soluble fraction of both lignins was used for melt spinning. The methanol soluble lignin fraction was able to continuously spin fibers free of defects, which is a significant improvement over the raw lignin. The methanol/methylene chloride fractions from both lignin demonstrate good solubility in electrospinning solution and performed very well in electrospinning process. Continus meltspinning for about 40 min was achived with both lignins and meltspinnning was performed for about 8 h. Carbon fibers were evaluated by SEM and tensile properties of fibers were measured. Diameter of melt spun carbon fibers were in the range of 15-30 Î¼m; while average diameter of elctrospun carbon fibers were in the range of 350-700 Î¼m, depend of concentration of electrospinning solution. Based on tensile properties data tensile strnght of meltspun switchgrass carbon fibers were 378 MPa and it was 441 MPa for meltspun softwood kraft carbon fibers. Results indicate lignin fractionation can provide different fractions with different characteristics and applications and can play a role like refinery for oil industry.

Copyright © American Institute of Chemical Engineers. All rights reserved.

465028 Chemically Modified Lignin Surfactant: Synthesis, Characterization and Their O/W Interfacial Properties

Abstract:

Lignin-based surfactant was prepared through chemical modification. This bio-based surfactant demonstrated to be a good emulsifier for W/O emulsion. Herein, grafting of butyric anhydride (BA), 3-(trimethoxysilyl)propyl methacrylate (MPS) and 2-bromoisobutyryl bromide (BIBB) onto kraft lignin to change its amphiphilicity was conducted. The chemical reactions were confirmed by FTIR, ^{1}H NMR, GPC, TGA analysis, and ^{31}P NMR. Lignin modified with MPS and BIBB could only dissolve partially in organic monomer due to the low grafting ratio. In contrast, over 95% of the hydroxyl group conversion was achieved when butyric anhydride was used. The butyric anhydride modified lignin resultant (lignin-B) was found to be completely soluble in oil monomers. According to the interfacial tension measurement, lignin-B was able to decrease the interface energy between water and styrene from 35 $J{\cdot}m^{-2}$ to 15 $J{\cdot}m^{-2}$. As expected, stable water-in-oil emulsion was prepared with lignin-B as emulsifier. The average diameters of emulsion droplets were 499.4 nm and 363.9 nm with a lignin-B content of 0.5 wt % and 5 wt % respectively. No separated aqueous phase was observed for 30 days. These results suggest lignin-B has promising potential to be used as a bio-based surfactant or emulsifier.

Copyright © American Institute of Chemical Engineers. All rights reserved.

466557 Reactive Diluents Prepared from Lignin Model Compounds:the Effect of Structure and Impurites on the Properties of Vinyl Ester Resins

Abstract:

Vinyl ester (VE) resins are utilized to produce polymer matrix composites for use in a wide range of applications due to possessing relatively high moduli, strengths, and glass transition temperatures (T_gs) while maintaining low weight and cost. VE resins often contain high concentrations of a petroleum-based reactive diluent (RD), such as styrene. In addition to lowering the resin viscosity, styrene is able to improve the overall polymer performance by increasing monomeric reactivity and extent of cure; however, styrene has been designated as a HAP, potential carcinogen, and a VOC. Renewable RDs with similar performance to petroleum-based RDs are desired to mitigate the aforementioned hazards as well as to facilitate the transition from petrochemical feedstocks to bio-based feedstocks.

Lignin is an abundant renewable resource with a high aromatic content that has the potential for the development of VE resins, including RDs. RDs that can be derived from lignin, including phenyl methacrylate (PM), 2-methoxyphenyl methacrylate (MG), and 4-propyl-2-methoxyphenyl methacrylate (M4PG), were synthesized and investigated as potential styrene replacements.Â A commercial VER was blended with each lignin-derived RD, and styrene, in 7:3 and 1:1 weight ratios and cured to â‰¥ 94 %, based on near-IR spectral analysis.Â The substituents on the aromatic ring of the lignin-derived RD are shown to have small effects on viscosity, thermogravimetric, and thermomechanical properties of the cured resins, where, in general, the smaller the RD, the higher the T_g and the lower the viscosity.Â Increasing purity of the RDs through chromatographic separations substantially increases the T_g and degradation temperatures of the resins. The thermogravimetric and thermomechanical properties of all lignin-derived RDs are directly comparable to that of styrene.Â These results suggest that PM, MG, and M4PG are effective bio-based RD replacements for styrene.

Copyright © American Institute of Chemical Engineers. All rights reserved.

472760 Fine Fractionation of Lignin By Molecular Weight Using Supercritical Fluids

Abstract:

As lignin is the only major constituent of biomass with aromaticity, it has been proposed as a renewable biopolymer replacement for petroleum-derived applications such as the polyols in polyurethane foams, the phenols in phenol–formaldehyde resins, and the polyacrylonitrile in carbon fibers. However, increasing evidence indicates that the molecular weight of the lignin should best be "tuned" for a given product application. In this study, concentrated acetic acid–water mixtures were used to dissolve most of a low-ash (1-2%) lignin recovered from a kraft black liquor via the SLRP process. Supercritical carbon dioxide (CO_2) was then added as an anti-solvent to the acetic acid–water–lignin mixture in increments of increasing pressure. Both the molecular weight and metals impurities of the lignin fractions that precipitated from solution were found to be a strong function of the pressure of the supercritical CO_2 being added to the acetic acid–water mixtures. The precipitated lignin fractions were characterized by GPC, ^{1}H, ^{13}C, and HSQC-NMR, and unexpected relationships between molecular weight and functional-group composition were obtained. Based on these limited results, we conclude that supercritical CO_2 can be an energy-efficient, alternative method for fractionating lignin in terms of molecular weight, while holding other molecular properties constant.

Copyright © American Institute of Chemical Engineers. All rights reserved.

460301 Ultralight, Reusable Cellulose Diacetate Aerogels for Selective Fluid Sorption

Abstract:

Uncontrolled oil or chemical spillage into fresh water can trigger a cascade of events including toxic loading and contamination of food resources, all of which are heavily detrimental to the ecosystem[1,2]. Widespread technologies proposed to deal with oil spills include oil skimming, *in-situ* burning, mechanical containment and utilization of dispersants, solidifiers and degrading microorganisms[3–8]. However, these methods are either inefficient or environmentally unfriendly. The need for efficient oil spill cleaning technologies was imminent in 2010 deep water horizon spill, one of the largest man-made disasters in the recent history, where oil gushed out in the Gulf of Mexico for straight 87 days, pouring out over 200 million gallons of crude oil[9]. In this type of scenario, the use of sorbents is attractive as they are easy to deploy and do not generate byproducts. There are three categories of sorbent materials: organic (milkweed, wood chips, rice straw, among others), inorganic (organo-clay, perlite and sand, zeolites, etc.) and synthetic (non-woven polypropylene mats, for example).[10–12] However, they (1) suffer from low sorption capability (typically less than 10 g per gram of solid support); (2) are difficult to recover (they are not buoyant), and (3) are not biodegradable.

In recent years, aerogels have gained attention as sorbent materials because of their light weight and high pore volume. In this study, highly porous (99.7% air volume) and ultra-light (4.3 mg/ml density) cellulose ester aerogels were synthesized for unprecedented water uptake (45-90 g/g) while affording wet strength and mechanical robustness (maximum compressive stress and strain of 350 kPa and 92%, respectively). The high compression strains are generally achieved with carbon nanofiber aerogels but the 92% strain is higher than those of nanocellulose aerogel reported in literature. The compressive stress of 350 kPa is 100 times higher than the reported cellulosic aerogels (see Figure 1). The aerogels were further adjusted towards hydrophobicity and oleophilicity via chemical vapor deposition with an organo-silane species. The as-modified aerogel exhibited high oil retention (20-30 g/g aerogel) while maintaining mechanical integrity for fast oil cleanup from aqueous media under marine conditions.

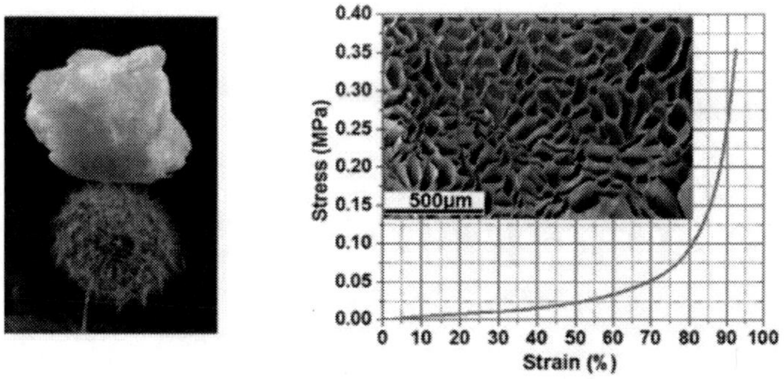

a) b)

Figure 1: a) Image of a 2 w/v% aerogel (density 4.3 mg/ml) on a dandelion leaf; b) Compressive stress-strain profile for 4 w/v% aerogels with the maximum compressive stress of 350 kPa and maximum strain of 92%. Inset: SEM image showing the radial cross-section of 4 w/v% CDA aerogel with a "honeycomb" like structure.

References:

1. Aguilera F, Méndez J, Pásaroa E, Laffona B. Review on the effects of exposure to spilled oils on human health. *J Appl Toxicol.* 2010;30(4):291-301. doi:10.1002/jat.1521.

2. Burton NHK, Musgrove a. J, Rehfisch MM, Clark N a. Birds of the Severn Estuary and Bristol Channel: Their current status and key environmental issues. *Mar Pollut Bull.* 2010;61(1-3):115-123. doi:10.1016/j.marpolbul.2009.12.018.

3. Fingas M. Oil Spill Science and Technology. *Elsevier, Amsterdam.* 2011;a.

4. Fingas M. "In-situ burning" in Oil Spill Science and Technology. In: *Elsevier*. Gulf publishing company, New York NY; 2011:737-903.

5. Michel J, Adams EE, Addassi Y, et al. *Oil Spill Dispersants. Efficacy and Effects.* Washington, DC; 2005.

6. Rosales PI, Suidan MT, Venosa AD. A laboratory screening study on the use of solidifiers as a response tool to remove crude oil slicks on seawater. *Chemosphere.* 2010;80(4):389-395. doi:10.1016/j.chemosphere.2010.04.036.

7. Correa M, Padron E, Petkoff I. The San Rafael De Laya oil spill : A Case of cleanup and remediation in Venezuela. In: *International Oil Spill Conference.*; 1997:932-935.

8. Aldhous P, Hecht J. Beware long-term damage when cleaning up oil spills. *New Sci.* 2010;10(206):2765. doi:10.1016/S0262-4079(10)61464-9.

9. Graham B, Reilly WK, Beinecke F, Boesch DF, Garcia TD, Murray C a. *Deep Water: The Gulf Oil Disaster and the Future of Offshore Drilling.*; 2011.

10. Teas C, Kalligeros S, Zanikos F, Stournas S, Lois E, Anastopoulos G. Investigation of the effectiveness of absorbent materials in oil spills clean up. *Desalination.* 2001;140(3):259-264. doi:10.1016/S0011-9164(01)00375-7.

11. Adebajo MO, Frost RL, Kloprogge JT, Carmody O, Kokot S. Porous Materials for Oil Spill Cleanup : A Review of Synthesis. *J Porous Mater.* 2003;(10):159-170. doi:10.1023/A:1027484117065.

12. Carmody O, Frost R, Xi Y, Kokot S. Adsorption of hydrocarbons on organo-clays-Implications for oil spill remediation. *J Colloid Interface Sci.* 2007;305(1):17-24. doi:10.1016/j.jcis.2006.09.032.

Copyright © American Institute of Chemical Engineers. All rights reserved.

464405 Exploring Cellulose Nanofibrils As Renewable Resource for Metal-Free Heteroatoms-Doped Carbon Electrocatalyst

Abstract:

With increasing development toward sustainable energy technologies for the future, a cost-effective electrochemical reaction has become a critical aspect for its practical applications. Replacing traditional metal-based electrocatalysts with carbon-based electrocatalyst offers both economical and environmental benefits. Cellulose nanofibrils, as biomass resources, have drawn a great interest as promising raw materials for preparing carbonaceous materials. In addition, enhanced electrochemical properties of carbon materials could be achieved by incorporating heteroatoms dopants. In this work, metal-free heteroatoms-doped carbons derived from carbon nanofibrils are explored as electrocatalyst for oxygen reduction reaction. The characteristics of doped carbons were investigated and their electrocatalytic activities were demonstrated.

Copyright © American Institute of Chemical Engineers. All rights reserved.

459777 Cellulose Nanofibers from Recycled Pulp: Production, Characterization and Application to Reinforce Recycled Paper

Abstract:

Papermaking is a sustainable sector where paper recycling has been recognized of a great importance [1, 2]. In Europe, 54% of the paper industryâ€™s raw material comes from recovered paper which corresponds to a paper recycling rate of 72% [3]. However, one of the main problems in the utilization of secondary fibers is to maintain the quality levels of the products that is continuously increasing following customer demands. On the other hand, paper demand has decreased due to both economic recession and replacement of paper by other information supports, causing cost pressures in the paper industry. Although natural and synthetic strength additives are commonly used in recycled paper, poor tensile strength is still the main source of customer complaints to paper manufacturers. In this context, new strategies to improve interfiber bonding must be explored and the use of cellulose nanofibers (CNF) are alternative additives to increase mechanical properties of paper with some advantages, such as their renewable nature, biodegradability and potential high availability.

CNF have gained increasing attention due to their high strength and stiffness combined with their low weight. These properties make CNF promising for applications in fields such as papermaking, composites, packaging, electronic devices, coatings, biomedicine and automotive. Among all of the applications, their use in the papermaking industry can improve paper quality and many studies have proved that the addition of CNF in mass to the pulp suspension increases the mechanical properties of paper [4]. CNF are currently produced from wood, mainly from eucalyptus one, but the reality is that the CNF market is still very limited [5]. The main drawbacks to use CNF in large volume industrial sectors, such us papermaking, is the cost of the product due to the high amount of energy consumption during CNF processing, the difficulties in producing uniform nanocellulosic particles, the dewatering difficulties and the problems to pump at high concentration [6]. A potential solution is to produce the CNF on-site from the mill pulp to avoid drying and transport costs, as well as dilution and dispersion problems.

Therefore this paper focuses on the on-site production of CNF from a cellulose source available at the recycled paper mill and its direct application to reinforce the recycled paper.

The production of CNF was optimized by studying different TEMPO-mediated oxidation levels before the homogenization mechanical process. The obtained CNF were characterized and several doses were added to the recycled pulp to evaluate their effect in terms of paper strength enhancement.

CNF were obtained from recycled newsprint pulp by TEMPO-mediated oxidation using 2.5, 5, 10 and 15 mmol of NaClO per each gram of pulp [7]. Once the pulp was oxidized, a cleaning process was performed through filtration steps using distillate water to reach a pH value around 7. Finally, homogenization at 600 bar was carried out in a PANDA PLUS 2000 laboratory homogenizer manufactured by GEA Niro Soavi (Italy).

Characterization of the oxidized cellulose pulp was performed by measuring the amount of carboxilate groups, as an indicator of the oxidation degree achieved after TEMPO oxidation, by conductimetric titration. CNFs were characterized by determination of the nanofibrillation yield, the cationic demand, the transmittance and the polymerization degree. Nanofibrillation yield was measured by centrifugation of a diluted CNF suspension (0.1% wt) at 4500 g for 30 min in order to isolate the nanofibrillated fraction in the supernatant from the non-fibrillated fraction deposited in the sediment. Cationic demand was measured by colloidal titration of the diluted suspension at 0.05%, with 0.001 N polyDADMAC, using a particle charge detector, Mütek PCD04, manufactured by BTG Instruments GmbH (Herrsching, Germany). Transmittance of the CNF suspensions diluted at 0.1% were measured between 400 and 800 nm of wavelength using a Cary 50Conc UV-Visible spectrophotometer manufactured by Varian Australia PTI LTD. Polymerization degree was calculated from the limiting viscosity number (intrinsic viscosity) of CNF suspensions, which was determined following the international standard ISO5351/1 with cupriethylendiamine as solvent [8].

Pulps were prepared through disintegration of 20 g of dry recovered paper in 2000 mL of water, by using a Messmer pulp disintegrator (Mavis Engineering Ltd, London). The recovered paper with the correspondent amount of CNF was left to soak at least 24 h before disintegration to favor swelling. A three-component retention system was added to the pulp (1.25 mg/g of coagulant, 0.75 mg/g cationic polyacrylamide as flocculant and 1.7 mg/g hydrated bentonite clay based on industrial recommendations). The pulp was used to prepare handsheets with basis weight of 60 g/m^2 in a normalized handsheet former Rapid-Köthen (ISO 5269/2, DIN 54 358). These handsheets were physically and mechanically characterized by using an AUTOLINE 300 from Lorentzen & Wettre (Stockholm, Sweden).

Results showed that cheaper CNF can be prepared on-site from recycled paper through chemical pretreatment followed by homogeneization treatment. The amount of carboxilic groups of the oxidized pulp increased linearly with the TEMPO-mediated oxidation level from 0.09 to 1.04 mmoles/ g o.d. pulp, for non-oxidized and 15 mmoles NaClO/g o.d. TEMPO-mediated oxidized pulp, respectively. Maximum CNF yield was achieved with 15 mmoles NaClO/g o.d. up to 77%. Cationic demand and transmittance also increased with TEMPO-oxidation level. The highest oxidized CNF achieved a 29% increase of tensile strength index by adding 3 wt% of CNF to the recycled pulp.

ACKNOWLEDGEMENTS

The authors wish to thank the Economy and Competitiveness Ministry of Spain for the support of the project NANOSOLPAPELREC with reference CTQ2013-48090-C2-1-R.

REferences

1. Miranda, R., Bobu, E., Grossmann, H., Stawicki, B. Blanco, A., 2010. "Factors influencing a higher use of recovered paper in the european paper industry". Cellulose Chemistry and Technology. 44(10), 419-430.

2. Blanco, A., Miranda, R. Monte, M.C., 2013. "Extending the limits of paper recycling: improvements along the paper value chain". Forest Systems. 22(3), 471-483.

3. CEPI, 2014. "Key Statistics. European Pulp and Paper Industry. http://www.cepi.org/topics/recycling".

4. Taipale, T., Osterberg, M., Nykanen, A., Ruokolainen, J. Laine, J., 2010. "Effect of microfibrillated cellulose and fines on the drainage of kraft pulp suspension and paper strength". Cellulose. 17(5), 1005-1020.

5. Chirayil, C.J., Mathew, L. Thomas, S., 2014. "Review of recent research in nano cellulose preparation from different lignocellulosic fibers". Reviews on Advanced Materials Science. 37(1-2), 20-28.

6. Osong, S.H., Norgren, S. Engstrand, P., 2016. "Processing of wood-based microfibrillated cellulose, and applications relating to papermaking: a review". Cellulose. 23(1), 93-123.

7. Saito, T., Kimura, S., Nishiyama, Y. Isogai, A., 2007. "Cellulose nanofibers prepared by TEMPO-mediated oxidation of native cellulose". Biomacromolecules. 8(8), 2485-2491.

8. Balea, A., Merayo, N., Fuente, E., Delgado-Aguilar, M., Mutje, P., Blanco, A. Negro, C., 2016. "Valorization of corn stalk by the production of cellulose nanofibers to improve recycled paper properties". Bioresources. 11(2), 3416-3431.

Copyright © American Institute of Chemical Engineers. All rights reserved.

472899 Phenolic Catalyzed Hydroxyl Radical Oxidation for Cellulose Nanofiber Production

Abstract:

Hydroxyl radicals (•OH) can change the surface chemical properties of cellulose to produce charged carboxylates in production of cellulose nanofibers (CNF). While ozone gas is an efficient oxidizing agent, ozone alone is relatively ineffective in oxidizing cellulose surfaces. We show that the addition of phenols from lignin and its residual byproducts are effective hydroxyl radical promoters in aqueous solution. Our preliminary results indicate that lignin-derived phenolics are effective catalysts to activate ozone for oxidation of the cellulose surface thereby improving cellulose dispersion in liquid media. Changes in the zeta-potential of the •OH radical modified CNF suggested a colloidal stability after the chemical treatment. The initial 7% yield of CNF from the starting never dried pulp feedstock increased after removing the surface modified CNF by centrifugation/filteration and re-applying the chemical treatment to the remaining cellulose. The •OH radical-cellulose reaction needs to be carefully controlled to inhibit cellulose structural degradation which affects CNF characteristics. We propose a model that predicts surface characteristics as a function of phenolic catalyst loading, ozone consumption, and mechanical treatment.

Copyright © American Institute of Chemical Engineers. All rights reserved.

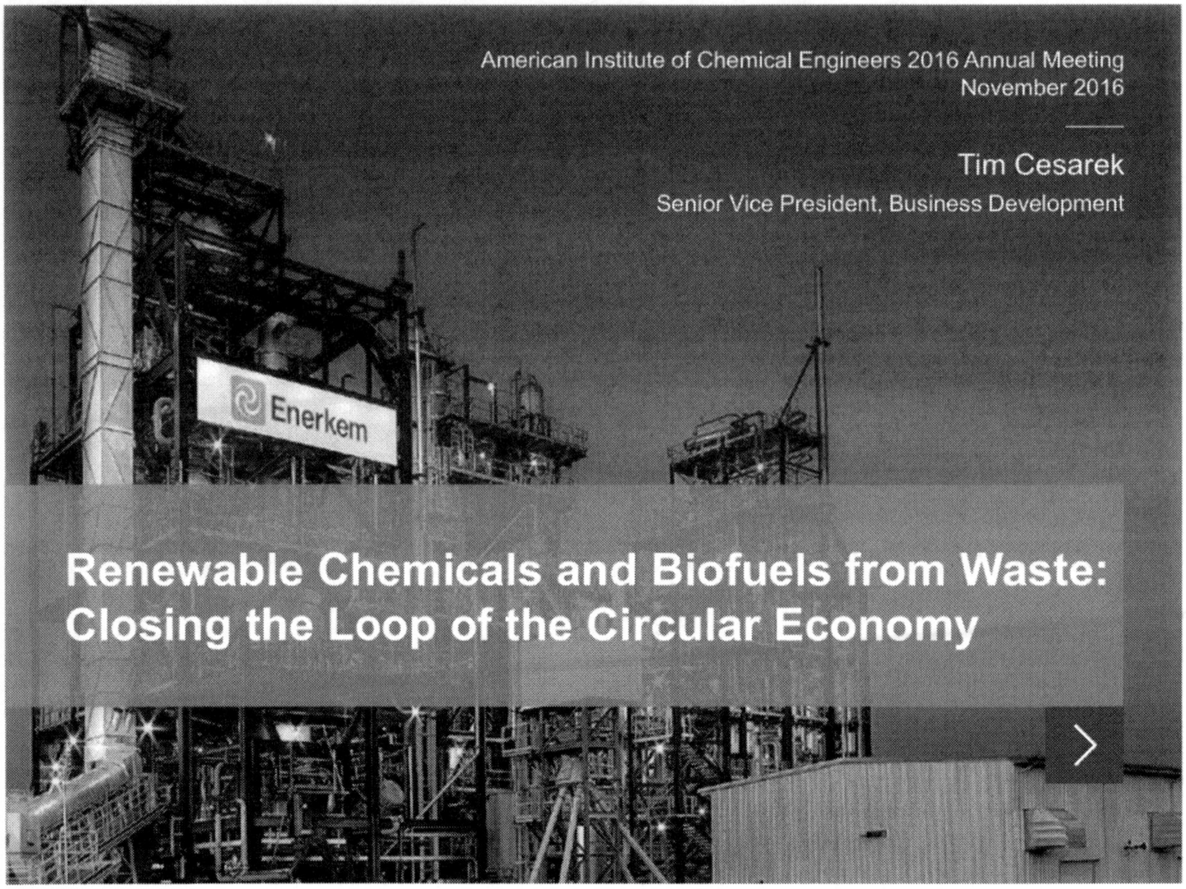

Enerkem at a glance

- MSW-based biofuels and renewable chemicals producer
- World's first full-scale commercial MSW biorefinery beginning operations in Edmonton, Alberta
- $400M invested to date to move from R&D to commercial stage
- 200 employees
- New alternative to landfilling and incineration

We're building the bioeconomy. | © Enerkem, 2016

Enerkem

Methanol sale and delivery

We're building the bioeconomy. | © Enerkem, 2016 Enerkem

Proprietary thermochemical technology and process

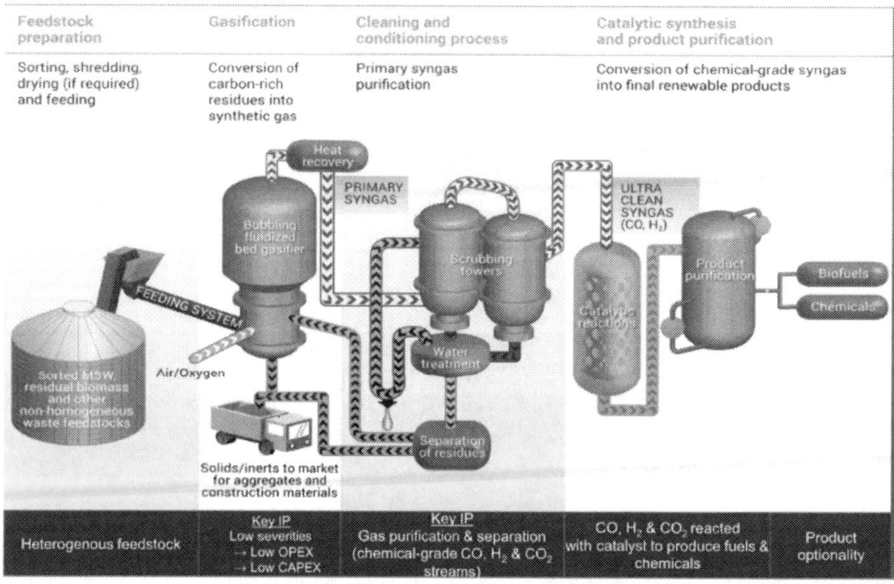

Low-severity technology capable of converting heterogeneous feedstock to clean syngas, the fundamental building block for all Enerkem's products

We're building the bioeconomy. | © Enerkem, 2016

Enerkem

Key market drivers for waste as feedstock

- Increased scarcity of urban landfill airspace and societal desire for waste diversion
- Circular economy or "cradle-to-cradle" approach
- Low cost unconventional feedstocks
- Renewable fuels mandates around the world
- Consumer pull for renewable and biobased products
- Focus on carbon footprint and GHG emissions reduction

We're building the bioeconomy. | © Enerkem, 2016

Enerkem

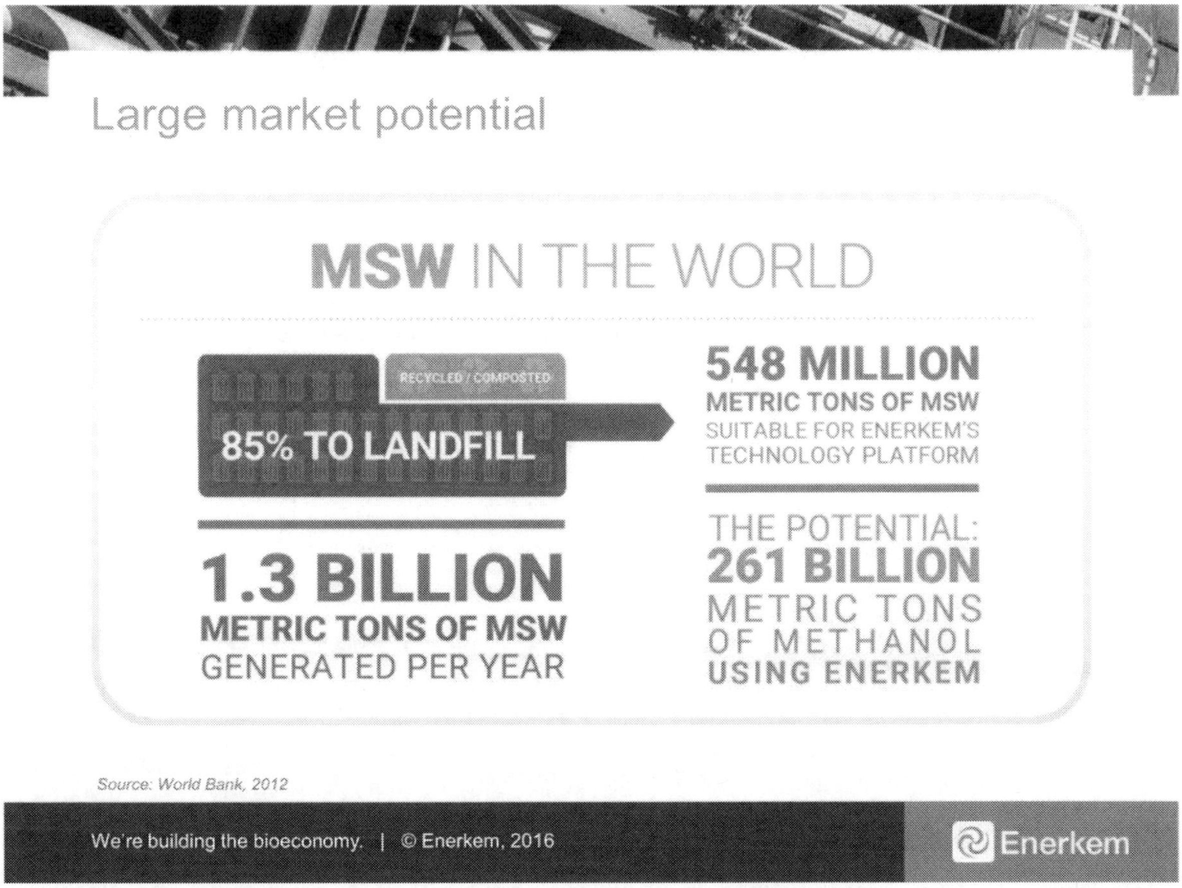
Large market potential
MSW IN THE WORLD
RECYCLED / COMPOSTED
85% TO LANDFILL
1.3 BILLION
METRIC TONS OF MSW
GENERATED PER YEAR
548 MILLION
METRIC TONS OF MSW
SUITABLE FOR ENERKEM'S
TECHNOLOGY PLATFORM
THE POTENTIAL:
261 BILLION
METRIC TONS
OF METHANOL
USING ENERKEM
Source: World Bank, 2012
We're building the bioeconomy. | © Enerkem, 2016
Enerkem

City of Edmonton's Integrated
Waste Management Centre
Edmonton
Recycled 20%
Composted 40%
Biofuels 30%
Landfill 10%
Waste diversion = 90%
1 Integrated Processing and Transfer Facility
2 Recycling center
3 Composting center
4 ENERKEM biorefinery
Alberta
Innovates
Energy and
Environment Solutions

VANERCO
Enerkem's second MSW to biofuel facility – Montreal Canada
Product: Biomethanol with ethanol expansion
Capacity: 45,000 MT; 69 million litres per year (18 million gallons) of biomethanol (1 train Enerkem plant)
Feedstock: 115,000 dry tons of RDF per year – C$20/MT RDF tipping fee
Location: Varennes industrial park (South Shore Montreal)
Engineering Start: Underway
Procurement Start: Q1 2017
Construction Start: Q3 2017
Operations Start: Q3 2018
Enerkem

Biofuels from waste increase energy diversity and reduce GHG emissions

Trends towards Low Carbon Transportation Fuels

Gasoline, Gasoline Blendstock or Replacement	Definition	Carbon Intensity (including land use effect)
CARBOB	CARBOB	99.78
Corn ethanol*	Derived from corn	75.97
Cellulosic Ethanol*	Derived from corn stover	41.05
Sugarcane Ethanol*	Derived from Brazilian sugarcane	56.66
Electricity	CA grid electricity	105.16
Hydrogen	Compressed H2 from central reforming of NG	151.01

** California Air Resources Board temporary CIs for fuels with indeterminate CIs*

Source: California ARB, July 31, 2015 (Proposed re-adoption of the LCFS: Third 15-Day Modified Regulation Order, Table 6. Tier 2 Lookup Table for Gasoline and Diesel and Fuels that Substitute for Gasoline and Diesel and Table 7. Temporary FPCs for Fuels with Indeterminate CIs)

We're building the bioeconomy. | © Enerkem, 2016 Enerkem

Using waste as a feedstock for the chemical industry

Waste-to-Chemicals public-private partnership in Europe

AkzoNobel
Enerkem
AIR LIQUIDE
van Gansewinkel
AVR.
Clean Tech Delta
Port of Rotterdam
Innovation Quarter
Gemeente Rotterdam
Provincie Zuid Holland
rotterdam partners

Location of Project in Port of Rotterdam

Product pathways of syngas and methanol

CO + H2 (syngas) and methanol are key building-block molecules of chemical industry

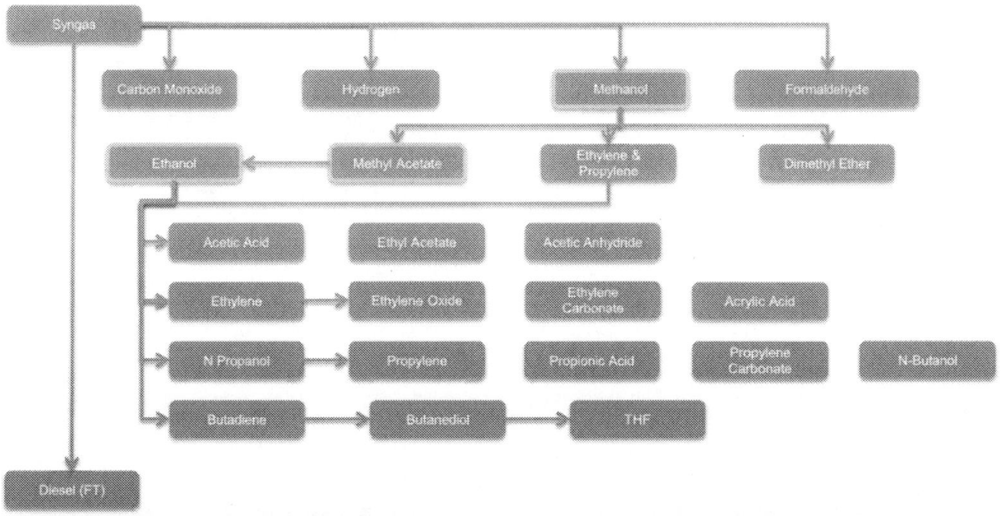

Multi-product platform being developed

Enerkem's platform is being expanded to new products. Acrylic Acid and Acetic Acid are the next steps:

Chemical building blocks in Enerkem's synthesis gas

CO

H_2

CO_2

Product	Global Market Size	Applications
Acrylic Acid	>$14 billion(1)	Industrial Coatings, Paints, Fiber
Acetic Acid	>$4.3 billion(2)	Solvents, Pesticides, and Coatings, Chemical intermediate
Ethyl Acetate	~$1.5 billion(2)	Coatings, Solvents, Ink
Butanol	>$2.4 billion (3)	Solvent, Chemical Intermediate, Fuel
Propanol	>$3.5 billion (3)	Fuel Additive, Plastic Intermediate, Solvents

Sources:
(1) ICIS. In USD
(2) ICIS, Company estimates
(3) Argus DeWitt

Target growth areas for global partnerships

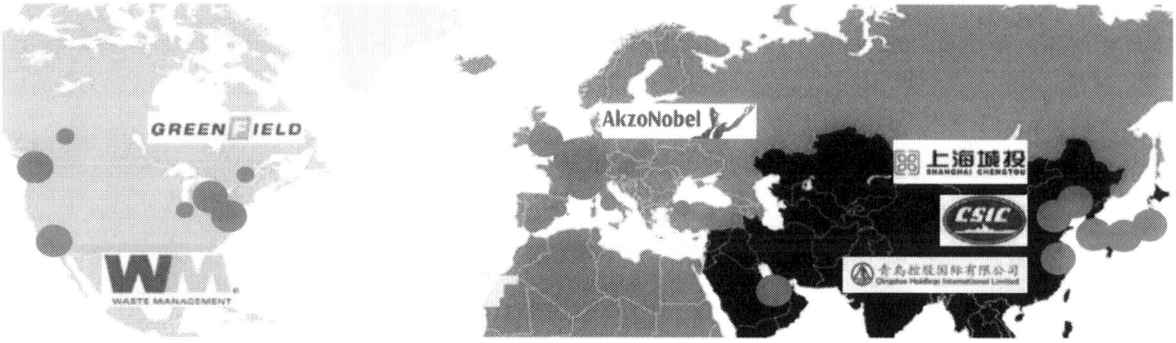

- Strategic partnerships with leading industrial groups
- Selection based on market attractiveness:
  - public policies
  - tipping fees
  - proximity to petrochemical infrastructure
  - population

We're building the bioeconomy. | © Enerkem, 2016

Enerkem

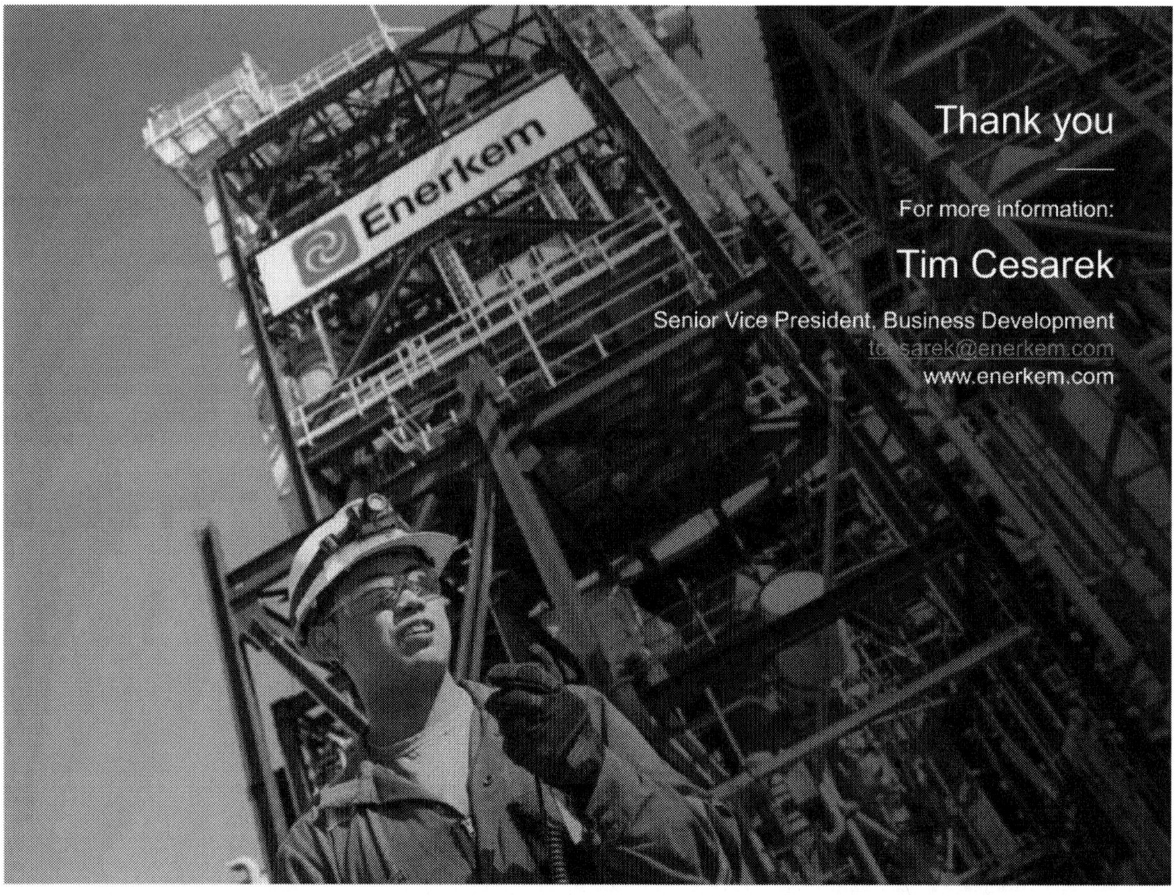

Thank you

For more information:

Tim Cesarek

Senior Vice President, Business Development

tcesarek@enerkem.com

www.enerkem.com

474725 Compact Steam Reformers for Power and Hydrogen Generation

Abstract:

Precision Combustion, Inc. (PCI) is developing high heat flux, compact, thermally-integrated steam reformer technology. The innovation consists of intimately coupled catalytic endothermic and catalytic exothermic sections implemented via novel metal mesh (MicrolithÂ®) supported catalyst. This results in high heat flux for improved product selectivity due to uniform temperature distribution, and flameless operation results in lower peak operating temperatures for increased material safety. These compact reformers are being implemented for fuel cell power generation and H_2 generation. Compared to conventional designs, PCIâ€™s approach offers improved performance, durability and power density. This is facilitated by the use of high surface area, low pressure drop MicrolithÂ® substrates coated with proprietary catalyst formulations. Additionally, implementation of flameless catalytic oxidation avoids high flame temperatures, increases thermal uniformity, distribution, durability and control. The burner component can utilize system waste heat, anode exhaust or supplemental fuel (as needed) to drive the endothermic steam reformer for efficiency advantages. Performance of the steam reformer has been rigorously evaluated for hundreds of hours with conventional as well as renewable fuels, including at high pressures for enabling hydrogen separation and purification. Data from these tests will be presented and discussed, demonstrating stable, equilibrium limited performance without coking or catalyst degradation. The sulfur tolerance studies of the burner and the steam reforming catalyst will also be presented.

PCI has extensive experience with balance of plant components and controls development. The BOP components e.g. fuel, air and water pumps and controllers are separately packaged, resulting in a compact system. The technology is scalable. The reformer was examined to generate SOFC-quality reformate and was coupled with a water-gas-shift reactor to generate HT-PEM, or PEM-quality (with additional CO cleanup) reformate. Additionally, integration with a hydrogen membrane separator resulted in production of high purity hydrogen. Alternatively, it could also be coupled with a pressure swing adsorber. Data from the reformer as well as from integrated testing with H_2 separator operating on HRJ and S8 will be presented, demonstrating the potential for a compact H_2 generation system. These reformers offer a solution towards compact, fuel flexible power generation systems for various military and industrial applications, as well as for on-site H_2 generation systems.

Copyright © American Institute of Chemical Engineers. All rights reserved.

458513 Development of Smart Ni Based Nano-Oxyhydrides for Hydrogen Production from Bioethanol

Abstract:

Development of smart Ni based nano-oxyhydrides for hydrogen production from bioethanol

L. Jalowiecki-Duhamel[1,*], Y. Romani,[1] W. Fang[1,2], C. Pirez[1], S. Paul[1], H. Jobic[1,3], F. Dumeignil[1]

[1]*Univ. Lille, CNRS, Centrale Lille, ENSCL, Univ. Artois, UMR 8181 - UCCS - Unité de Catalyse et Chimie du Solide, Lille, 59000, France*
E-mail: louise.duhamel@univ-lille1.fr
[2]*Key Laboratory of Medicinal Chemistry for Natural Resource - Ministry of Education, School of Chemical Science and Technology, Yunnan University, Kunming, 650091, China*
[3]*IRCELyon Institut de Recherches sur la Catalyse et l'Environnement de Lyon, 69626 Villeurbanne Cedex, France.*

At the world level, people are concerned by their health and even life degraded by environmental pollution. Therefore, the potential benefits of a hydrogen economy coming from renewable resources are creating a large consensus and increased attention is focused on hydrogen production technologies. So far, hydrogen has been produced from ethanol by steam reforming, partial oxidation, oxidative steam reforming, and autothermal reforming reactions. The endothermic steam reforming reaction extracts more hydrogen atoms from ethanol and water, however, additional energy supply is needed. One alternative way of supplying heat is to add oxygen or air to the feedstock and simultaneously to burn a portion of ethanol. [1-3] Moreover, in complement, some chemical energy can be also brought by a "smart" catalyst, such as an oxyhydride. [4-7]

Well tuning the preparation, formulation and conditions applied, different series of nickel based nano-oxyhydride (Ni-Ce-(Al)-O and Ni-Mg-Al-O) catalysts were developed for the highly efficient and sustainable H_2 formation from ethanol in the presence of oxygen and water with low energy input. The chemical energy released from the reaction between hydride species stored in the nano-materials and O_2, is used to obtain continuous complete conversion of ethanol with simultaneously production of H_2 with an oven temperature at only 50°C (Figs 1 and 2). The influence of different parameters on the activity and selectivity was analyzed, such as the reaction temperature, water-oxygen-ethanol feed compositions, as well as Ni content and in-situ pre-treatment in H_2 of the catalysts. Moreover, different physico-chemical characterizations were performed allowing a discussion on active site and mechanism.

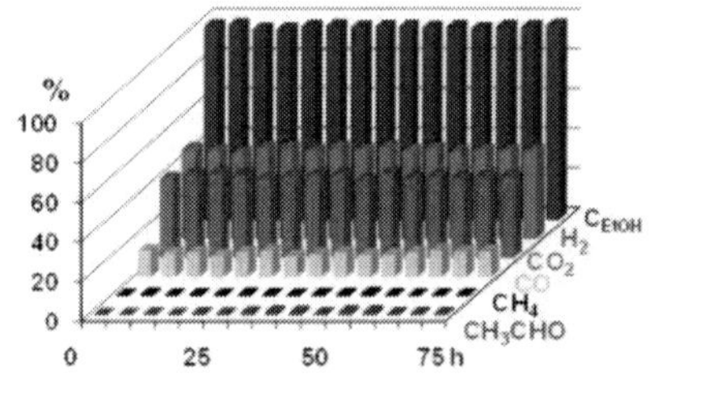

Fig. 1: Ethanol conversion and products distribution on Ni-Ce-O based nano-oxyhydride.

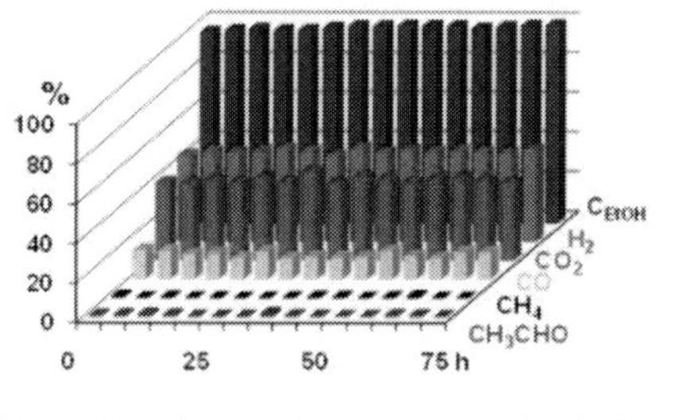

Fig. 2: Ethanol conversion and products distribution on Ni-Mg-Al-O based nano-oxyhydride.

1. G. A. Deluga, J. R. Salge, L. D. Schmidt, X. E. Verykios, *Science* **30**, 5660, 993 (2004).
2. V. Subramani, C. Song, *Catalysis* **20**, 65 (2007).
3. L. V. Mattos, G. Jacobs, B. H. Davis, F. B. Noronha, *Chem. Rev.* **112**, 4094 (2012).
4. W. Fang, C. Pirez, S. Paul, M. Capron, H. Jobic, F. Dumeignil, L. Jalowiecki-Duhamel, *ChemCatChem* **5**, 2207 (2013).
5. C. Pirez, M. Capron, H. Jobic, F. Dumeignil, L. Jalowiecki-Duhamel, *Angew. Chem. Int. Ed* **50**, 10193 (2011).
6. C. Pirez, W. Fang, M. Capron, S. Paul, H. Jobic, F. Dumeignil, L. Jalowiecki-Duhamel, *Appl. Catal. A* **518**, 78 (2016).
7. W. Fang, C. Pirez, S. Paul, M. Jiménez-Ruiz, H. Jobic, F. Dumeignil, L. Jalowiecki-Duhamel, *Int. J. Hydrogen Energy* (2016) DOI: 10.1016/j.ijhydene.2016.04.019.

- Abstract Text

Copyright © American Institute of Chemical Engineers. All rights reserved.

466889 Exploring Transition Metal Carbides and Phosphides for Ex-Situ Catalytic Fast Pyrolysis

Abstract:

Hexagonal Fe_2N type molybdenum carbide (Î²-Mo_2C) has been explored as biomass upgrading catalyst in recent decades for model compounds via hydrodesulfurization (HDS), hydrodenitrogenation (HDN), and hydrodeoxygenation (HDO). Molybdenum carbide has also been shown to be a superior catalyst support, for example, for platinum in comparison to carbon in electrocatalytic water splitting. However, there have not been any studies to investigate the effects of supporting biomass upgrading catalysts such as nickel phosphide (Ni_2P) on molybdenum carbide. Additional carbide-phosphide interfacial sites thus generated have the potential to be of tremendous importance in biomass conversion catalysis. One of the main stumbling blocks for biomass derived pyrolysis oils finding wider use in energy and chemical feedstock generation is the presence of significant amount of oxygen, unlike in fossil reserves. While model compound studies using phosphide and carbides have shown encouraging HDO activities, only a limited number of studies have explored the HDO catalysis on abundant biopolymers such as lignin and cellulose. This presentation will be based on the ex situ catalytic fast pyrolysis of lignin and cellulose in the presence of molybdenum carbide supported nickel phosphide. Particular emphasis will be devoted to the characterization of the interfacial sites and the effect of different Ni_2P loadings on catalysis. Stability of the carbide support during the synthesis of the composite material will also be discussed. In addition, a detailed analysis of product fragments that originate due to the presence of interfacial catalyst sites will be presented via comparison with catalysis results from Mo_2C and Ni_2P individually.

Copyright © American Institute of Chemical Engineers. All rights reserved.

469915 Microwave-Assisted Fast Pyrolysis (MW-FP) of Several Lignocellulosic Feedstocks

Abstract:

Using fast pyrolysis processes, renewable biomass resources can be converted to bio-oils that retain more than 60% of the original energy content, and these oils can then be upgraded to motor fuels or other commodity chemicals. In order to attain economic feasibility, feedstocks for commercial-scale fast pyrolysis must be obtained from a diverse set of low-cost resources for which the conversion performance is well-understood. This presentation summarizes the development of a microwave-assisted fast pyrolysis (MW-FP) reactor that employs a 3 kW microwave beam to heat samples of 1 to 2 grams at rates as fast as 200 °C/s. Liquid, solid and gas yields from as many as 8 samples per day can be measured and analyzed to determine product qualities. Because the individual experiments can be conducted in minutes rather than hours and only small quantities of material are required, MW-FP is advantageous for parametric studies exploring the effects of feedstock material, conversion temperature, and heating rate on fast pyrolysis product yields. Direct measures of product yields show that oil yield increases with volatile matter and lignin contents and decreases with potassium and sodium contents. Contents of total ash, calcium and magnesium did not appear to measurably affect oil yield on a dry, ash-free material basis. Results from the MW-FP reactor have been calibrated to floor-scale fluidized bed fast pyrolysis tests to ensure reliability of the conclusions.

Copyright © American Institute of Chemical Engineers. All rights reserved.

Biomass to Energy – A novel process

By Rakesh Gupta
Lead Process Engineer, Lockheed Martin Energy
Lockheed Martin Corporation
Moorestown, NJ

Background

Biomass is defined as organic material derived from living organisms especially plants. It is an abundant source of renewable energy and chemicals. It is widely available in the United States and worldwide. Its use is being encouraged worldwide as a source of energy, especially in the area of power production. Several government programs have initiated subsidy programs to help develop the power generation from renewable sources. Since biomass is a renewable source, it can be a significant source of fuel for power generation.

Wood is the largest source of biomass. It can come from crop residues, forest thinning, milling residues, agricultural process residues and even municipal solid waste (MSW).

Biomass constituents include organics such as cellulose and lignin. They also contain water and other components such as chlorine, nitrogen, phosphorus, sulfur, silicon and trace metals.

The most important characteristic of a biomass is its heating value. This value is reduced by the presence of water and ash. Although, water presence in the biomass can also be beneficial by assisting in the steam reformation process discussed later.

Lockheed Martin is committed to developing renewable energy sources and has committed significant resources towards its development. Lockheed Martin has partnered with Concord Blue, licensor of a novel advanced gasification technology to convert biomass to energy. The purpose of this partnership is to develop a scalable and fully integrated solution. The Concord Blue process utilizes a patented method of “staged” reforming to convert organic material into a hydrogen rich syngas. This syngas is cleaned of contaminants and it can be either separated into valuable chemicals or fed to a combustion engine or turbine for power generation.

These power plants can range from 250 kW up to 20 MW in size.

Technology

There are many ways to use biomass as a source of energy for power generation. The most common method is to use it as a co-fuel with other sources such as coal, natural gas, etc. However, biomass can also be gasified using air or oxygen or pyrolyzed using a heat source. Some systems have a combination of pyrolysis and combustion/oxidation processes.

Syngas, short for synthetic gas, is the generation of gas from an organic fuel such as MSW, or woodchips by pyrolysis. In this process, the organic waste is gasified by addition of heat. The gases produced have a higher hydrogen to carbon ratio than the source material. The gases also contain carbon monoxide, hydrogen, methane, tars and other contaminants.

Pyrolysis involves the decomposition of organic compounds into gases and solids. This is done in the absence of oxygen. When the biomass is initially fed to the pyrolyzer, the volatile components are released almost immediately. The remaining char is converted to H_2 and CO in slower reactions.

Steam reformation process involves converting the volatile components such as methane and other organics into hydrogen with endothermic and reversible reactions. This process requires addition of steam and energy. These reactions can be repeated for higher hydrocarbon chains also, although each of those reactions have their own energy requirements.

In the Concord Blue staged steam reformation process, syngas is generated with components such as hydrogen, methane, carbon monoxide and carbon dioxide along with other tars and contaminants.

The raw syngas is cooled and cleaned in a scrubber system and then fed to a combustion engine to generate power, but ultimately the syngas could have other uses as well. The components of the syngas could be separated into valuable chemicals or converted into biofuels.

The process generates many valuable byproducts such as char that may be used for agricultural applications, steam for plant use and heat that could be used for applications such as hot water for district heating. Biochar produced from the pyrolysis process can also be used as a source of fuel for the pyrolysis process. It can also be used as activated carbon feedstock.

Syngas generated from the conversion process can also be converted into synthetic crude oil which can be further refined into clean synthetic fuels and chemicals such as kerosene, diesel, jet fuel, lubricants, etc. It can also be blended into crude oil and processed in a refinery. The practice involves using the Fischer-Tropsch process.

Suppliers

Gasifiers are provided by many suppliers such as GE, Conoco, Shell, Siemens, MHI and KBR. Shell and Siemens systems are oxygen blown using slagging type burner. MHI and KBR technologies use air or oxygen, do not use a burner and are non-slagging type.

The technology licensed by Lockheed Martin from Concord Blue utilizes staged reforming as a method of generating synthetic gas. The organic material is initially pyrolyzed and then reformed using steam and heat.

Pilot Plant

In order to demonstrate this technology, Concord Blue's patented method of "staged reforming", Lockheed Martin has built a pilot scale project at its Owego, NY facility. Lockheed Martin selected wood chips as the initial fuel source, since they are easily obtained in consistent size, they are available in abundant supply and they provide a consistent source of organics. Literature search showed wood chip samples from New York state have heating values of 5,600 Btu/lb. to 7,800 Btu/lb., moisture content of 28 – 42 % and ash content from 0.4 to 1%.

The size of this of this project is 250 kWe. It has the capability of producing renewable / sustainable power using various solid wastes, organic wastes, non-organic residues and woody biomass feedstocks such as clean wood chips and refuse derived fuel (RDF), while showcasing the technology. The feedstock will initially be woody biomass with an average moisture content of 37%.

The following is a simple sketch of the Concord Blue technology.

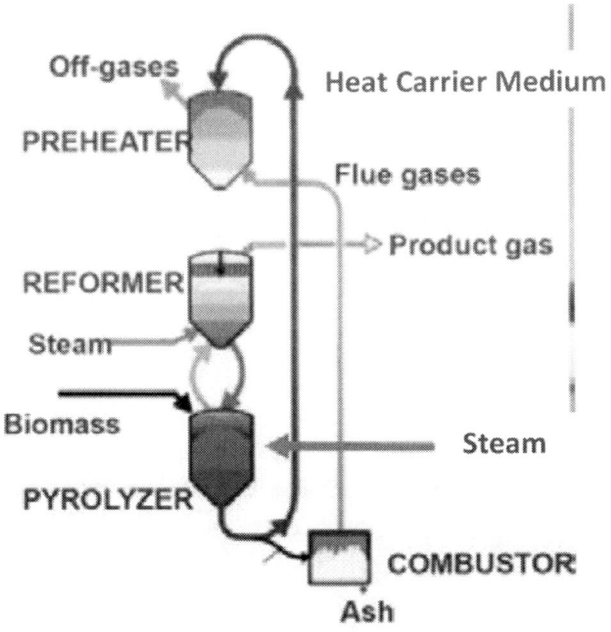

Figure 1 – Concord Blue Staged Reforming Technology
(Institute of Wood Technology and Wood Biology (HTB), Hamburg, Germany)

The biomass is fed to the pyrolyzer at temperatures above 900 °F. The gases generated are reformed in a steam reformer to increase the hydrogen concentration in the syngas. Heat is supplied to the reformer and the pyrolyzer by a heat carrier medium. The heat carrier medium is heated in the preheater as shown above in Figure 1.

The 250 kW Technical Assessment and Demonstration Facility's subsystems include:

- Feedstock Storage and Handling - The biomass is received in specially equipped trucks. It is transferred to the pyrolyzer using a number of conveyors. All feedstock materials are processed off-site by a feedstock provider and transported to the plant. Feedstock processing includes sorting, shredding to size, and removing all ferrous and non-ferrous metals as well as all inert material such as rocks, stones, ceramics, etc.

- Staged Reforming Equipment - The primary function of the Staged Reforming System is to gasify the incoming feedstock to generate syngas for downstream combustion in the Internal Combustion Engine. Biomass is fed to the Pyrolyzer where it gasified in a slow and controlled manner. Gas from the Pyrolyzer is fed to the Reformer where higher temperature reactions take place in the absence of oxygen. It is in this high temperature reformer where the water gas shift reaction occurs. It consumes water and energy to increase the heating value of the syngas by generating hydrogen. This is also where tars are cracked with water to produce hydrogen and carbon monoxide. This step improves the quality of the syngas by reducing the amount of tars that end up downstream into the syngas cleanup equipment potentially fouling piping and the gas cleaning system, thereby reducing gas flow to the generator

- Burner - The Burner is a vendor supplied package, that is designed to combust natural gas, syngas and other fuel sources, such as spent biodiesel, in order to generate hot flue gas to supply heat to Staged Reforming System's Pre-Heater. Biodiesel is atomized into the burner using compressed air. Fresh outside air, using combustion air blower provides the oxygen source for combustion within the burner. During start-up, natural gas is used to fuel the burner and bring the flue gas up to operating temperature. Note that in other installations, the char generated by the process will be combusted to provide the heat for the Staged Reforming System.

- Syngas Cleaning - Raw syngas from the Staged Reforming System Reformer contains a number of contaminants not suitable for running an IC engine, such as acid gases (i.e. H_2S, HCl, HF, COS), ammonia, particulate matter, and tars. (The presence and concentration of contaminants is affected by the type of fuel used.) The function of Syngas Processing and Conditioning system is to clean and condition the raw syngas from the Staged Reforming Process so that it conforms to the syngas quality requirements of the internal combustion (IC) engine. The first step in the syngas scrubbing process is the quenching of the hot gases. The gases then pass through a venturi scrubber to remove particulates. The system includes a scrubber to remove ammonia and another scrubber to remove H_2S. The syngas pressure is boosted to meet the requirements of the engine. Also, the gases are reheated to reduce the humidity of the gases.

- Flue Gas Cleaning will be added in the future depending upon initial results of environmental stack testing.

- Syngas Storage / IC Engine-Generator - The function of the Power Generation and Gas Storage system is to convert the cleaned syngas into electrical energy through the IC Engine. The gas storage system provides a mechanism to even out the composition of the syngas and provides a few minutes of storage to allow for graceful shutdown of the engine, when required. A single 250 kWe internal combustion engine vendor package is used to convert the cleaned syngas into electrical energy.

- Flare – The flare is used to burn off any off-specification syngas or cleaned syngas in case the engine is not operating. The flare provides safe combustion of the syngas to prevent harmful toxins and combustible materials from emitting to the atmosphere. The flare is designed to maintain a constant flame at the flame tip. Nitrogen is used as the purge gas for the flare velocity seal.

- Wastewater Treatment – The wastewater treatment system collects the water discharge from the scrubbing system. Currently, it is shipped offsite for treatment. In other applications, it would be treated on-site and returned to the plant for process water uses.

- Cooling Water System – The quenched gases in the syngas scrubbing system require cooling for the engine to combust them. They are cooled using a closed circuit cooling tower.

- Compressed and Instrument Air System – The compressed air system provide the air required for the operation of the instruments, valves and the atomizing air for the burner.

- Natural Gas – Natural gas is currently used as the fuel source for heating the heat carrier medium.

- Nitrogen – Nitrogen is used in the process to purge out the system before generating syngas. It is also used in a number of components to prevent mixing of syngas with air.

- Biodiesel – Biodiesel is used as the absorbent for the tars generated in the process. The spent biodiesel is used as a fuel in the burner.

A simplified flow diagram of the system is as follows:

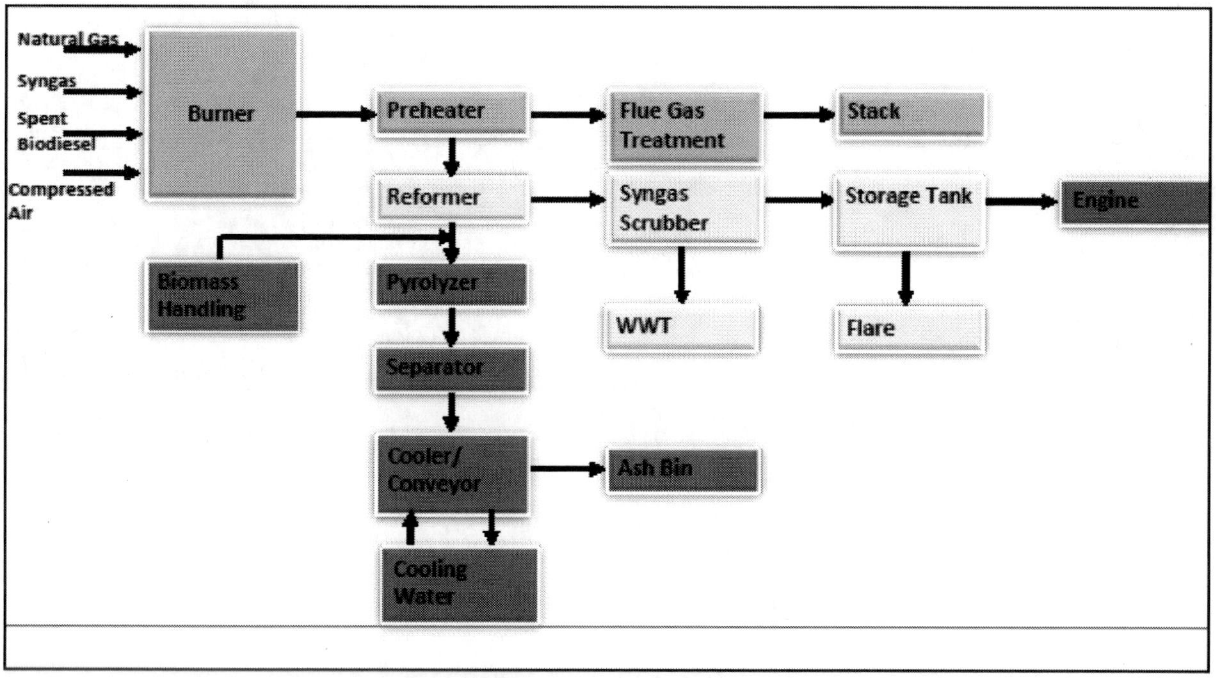

Figure 2 - Block Flow Diagram of the Owego pilot plant

The following is conceptual design of the pilot plant:

Figure 3 – Initial concept of the pilot plant

The following Figure 4 is a picture of the installed system.

Figure 4 – Owego pilot plant

Commercial Scale Plant –Herten, Germany

This is a full scale plant that will be built to generate more than 5 MW of power using wood chips as a fuel. It is similar to the pilot plant described above, except fuel drying has been added to the process.

Other energy-saving strategies are being implemented in this design, including use of biochar as a heat source for the Staged Reforming System, and use of heat from engine exhaust as a heat source for the biomass dryer.

The pictures below Figures 5 & 6 show a conceptual view of the original site plan.

Figure 5 – Herten Biomass Plant Conceptual View

Figure 6 – Herten original site plan

Results

The Owego pilot plant has completed the subsystems commissioning. System level commissioning underway. In the last month, we have made syngas that meets the quality requirements of the engine and have run the engine briefly. Plans are under way to complete the tuning and optimization of the process. Complete results of the tests will be available later this year.

The Herten, Germany project is still under design and will begin construction later this fall.

References

1. "Elemental Analysis of Wood Fuels", Final Report, June 2013, NYSERDA Report 13-13

GENERATION OF HYDROGEN FROM SOLID FEEDSTOCK USING THE HEATPIPE REFORMER TECHNOLOGY WITH IN-SITU HYDROGEN SEPARATION BY NICKEL MEMBRANES

Jonas M. Leimert, Jürgen Karl
Institute of Energy Process Engineering, Friedrich, Alexander-University Erlangen-Nuremberg, Germany

Introduction

The Heatpipe Reformer was developed as an allothermal gasification system at TU Munich during the EU project ``BioHPR" (Ref. ENK5-CT-2000-00311). The concept was commercialized by Agnion Inc., which built a demonstration plant in Pfaffenhofen, Germany in the scale of 500 kW and two commercial plants in the 1 MW scale in Grassau, Germany and Auer, Italy. Results from these plants are given by Gallmetzer in [1].

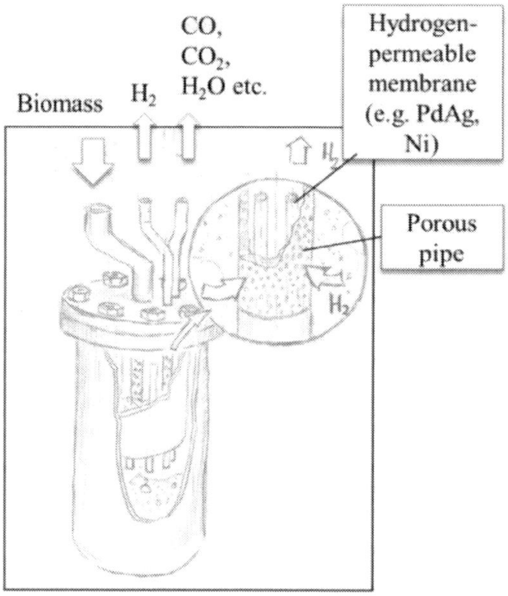

Figure 1: Sketch of Heatpipe Reformer with integrated hydrogen separation by membranes

Figure 1 shows a process scheme of the Heatpipe Reformer: In contrast to other dual fluidized bed gasification processes, the gasification takes place in a pressurized reformer chamber at 2-10 bar and 800°C. The gasification system can process both biomass and coal as feedstock. The pressurized synthesis gas allows usage for SNG synthesis, combustion in engines or gas turbines. Alternatively, the hydrogen in the synthesis gas can be separated for coupled heat and hydrogen generation. This could be achieved by a conventional process chain using a downstream CO shift reactor and hydrogen purification systems. As proposed by Karellas et al. and Schäfer the hydrogen could also be separated directly in the reformer chamber [2], [3]. This system enhances hydrogen generation due to the shift of the water-gas reaction to the product side caused by hydrogen removal:

$$CO + H_2O \rightleftharpoons CO_2 + H_2$$

For this application dense hydrogen permeable membranes with a solution-diffusion mechanism can be used, providing hydrogen with highest purity. Such dense hydrogen permeable membranes have been covered extensively over the years. In the case of palladium membranes the temperature regime of these investigations was in the range of 200-600°C restricted by the kinetics of hydrogen permeation at lower temperatures and the poor durability of Pd/Ag composite membranes at higher temperatures. Also membrane systems with other materials like Niobium, Tantalum, Titanium, Vanadium or Nickel have been proposed with varying limitations regarding temperature, pressure and atmospheres. For the Heatpipe Reformer at FAU-EVT the membrane system has to withstand temperatures of 750-900°C and pressures of up to 5 bar.

State-of-the-art

The approach at the Institute of Energy Process Engineering is the usage of dense nickel capillaries for hydrogen separation due to their low cost and high durability at the desired process temperature. Nickel membranes were already used in the Heatpipe Reformer system at FAU-EVT for hydrogen degassing of liquid metal heat pipes as described in [4]. In this application the nickel capillaries were operated for 2500 h at 800 – 870°C in sodium atmosphere without failure.

This section introduces the mathematical background of hydrogen permeation and the analysis of turnover rates and hydrogen yields. It also shows the state-of-the-art of hydrogen permeable membranes for high-temperature applications.

Hydrogen permeation and process analysis

The permeation of diatomic gases like hydrogen through a dense body is divided into several steps: The adsorption of the hydrogen on the surface, the dissociation into protons and dissolution into the metal, which is sometimes referred to as solution of the hydrogen. The hydrogen diffuses through the metal as protons, which is in most cases the rate-controlling step of the whole process, finally recombines and desorbs from the solid [3], [5], [6].

The following equation describes the hydrogen molar flow through a dense membrane n_{H2}, it depends on the permeability P of the membrane material, membrane area A, thickness Δx and the square root of the pressure difference between retentate and permeate p_1, p_2 respectively.

$$\dot{n}_{H_2} = -P \cdot \frac{A}{\Delta x} \cdot ({p_1}^{0.5} - {p_2}^{0.5}) \tag{1}$$

The permeability itself is a temperature activated material property and can be calculated with the Arrhenius equation using the permeability constant P_0 permeability activation energy E_A and gas constant R at the respective temperature T:

$$P = P_0 \cdot e^{-\frac{E_A}{R \cdot T}} \tag{2}$$

To describe the hydrogen separation in relation to the amount of treated gas, the permeate flux is divided by the maximum amount of hydrogen, which is possible to generate from the feed. The result is the dimensionless hydrogen separation rate Y_M.

$$Y_M = \frac{\dot{n}_M}{4\,\dot{n}_{CH4} + \dot{n}_{CO} + \dot{n}_{H2}} \tag{3}$$

The feedstock in gas phase reaction consists mainly of CO and CH_4. To describe the course of the reaction, the CO and CH_4 turnover X_{CH4} / X_{CO} is introduced:

$$X_i = \frac{\dot{n}_{i,out}}{\dot{n}_{i,feed}} \tag{4}$$

State-of-the-art of high temperature hydrogen permeable membranes

Figure 2 shows a summary of the literature study of high temperature membrane systems. Additionally to the permeability values of Pd based membranes mentioned above, also values from Akamatsu et al. [7], Hatlevik et al. [8], Yan et al. [9], Kusakabe et al. [10], Fernandez et al. [11], Guo et al. [12], Mardilovich [13] and Xomeritakis and Lin [14] are included. Palladium composite membranes show high hydrogen fluxes nearly reaching the physical boundary, which is shown by the hydrogen flux calculated from pure palladium. However, many membranes show fluxes well below this ideal value covering up to three orders of magnitude. The temperature limit for these membranes lies around 650°C making an application in gasification processes very challenging. Further problems of Pd-composite membranes in synthesis gas applications are coke formation, which can lead to membrane failure as carbon atoms diffuse into the Pd lattice leading to cracking [15]. They also show a degradation in hydrogen flow when exposed to synthesis gas components like CO, H_2O or sulfur components like H_2S [15]–[18].

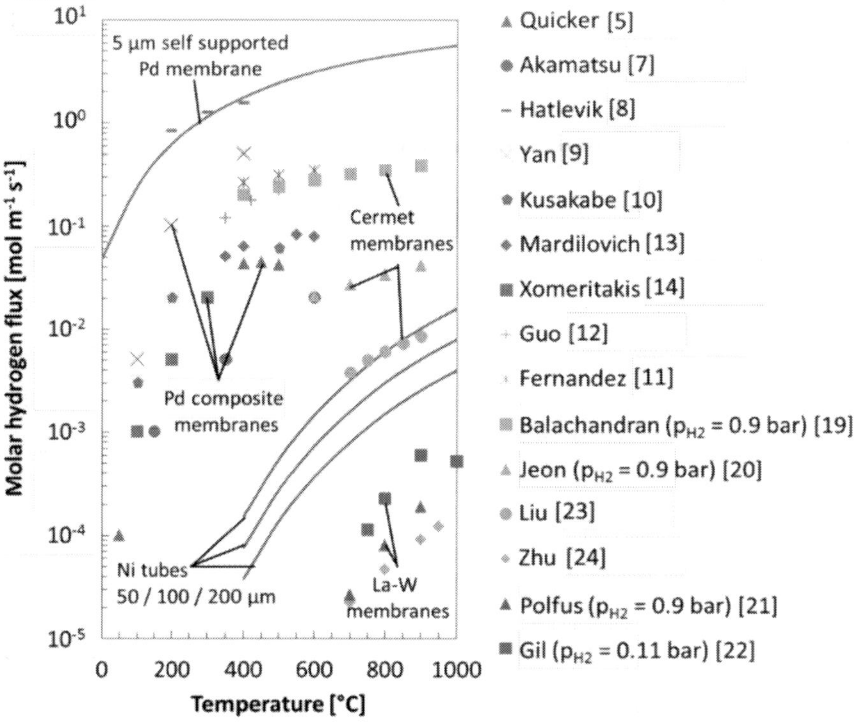

Figure 2: Literature values of membranes for high temperature hydrogen separation calculated from published permeability values at P_{H2} = 1 bar unless otherwise stated

Future membrane solutions could come from cermet or La-W membranes as described in [19]–[24], which combine high molar fluxes with temperature stability and tolerance for many gas components. However, the transfer from laboratory to an industrial application could again pose problems. Recent literature describes only small, disk shaped membranes.

Nickel membranes show in the temperature regime of gasification hydrogen fluxes in the same order of magnitude as some of the Pd and cermet membranes while showing high stability. Considering low nickel prices, this material could be easily used in process engineering without an elaborate membrane fabrication process. When small tube diameters are used, high membrane areas can be accomplished easily by constructing membrane bundles.

Gasification experiments

Fig. 3 shows a CAD sketch of the 100 kW Heatpipe Reformer at EVT-FAU. The reformer pressure vessel is integrated into the combustion chamber, described in a previous publication [25]. Eight heat pipes, filled with sodium, transfer the heat, which is needed for gasification, from the combustor into the reformer. The fluidized bed furnace is made of refractory concrete to withstand the temperatures and abrasive conditions of the fluidized bed. The primary air inlet is located in the lower part of the furnace. Before entering the combustion chamber the primary air is preheated. Therefore, the air is lead through an integrated double walled heat exchanger with a pillow-like structure located at the external wall of the freeboard. The secondary air is provided by four steel pipes with a nominal diameter of DN25 and is blown into the freeboard radially. The flue gas is led out of the combustor at the top with a temperature of about 200-400°C after preheating primary and secondary air.

The reformer is filled with olivine with a particle size of 0.063 - 0.180 mm from Sibelco Europe with a bed height of 1.1 m. Pressurized steam and lignite is fed into the gasifier through a stand pipe. This ensures a thorough mixing of steam, lignite and olivine. The synthesis gas leaves the gasifier with a temperature of about 500-700°C through an upper outlet.

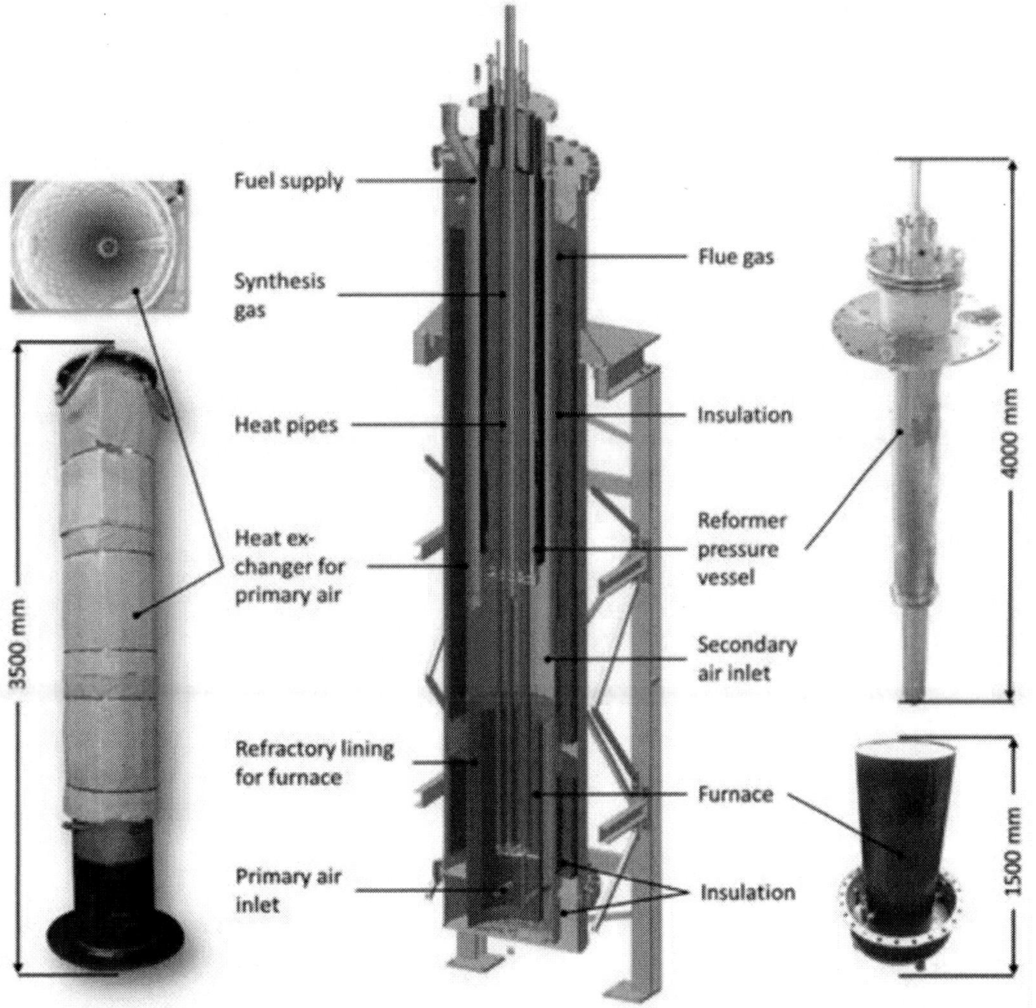

Figure 3: CAD sketch and photographies of the EVT Heatpipe Reformer

Experimental setup

The Heatpipe Reformer described above is integrated in a process chain which guarantees the fuel, air and steam supply as well as the conditioning and analysis of the synthesis gas. The piping and instrumentation diagram of this process chain is shown in Figure 4.

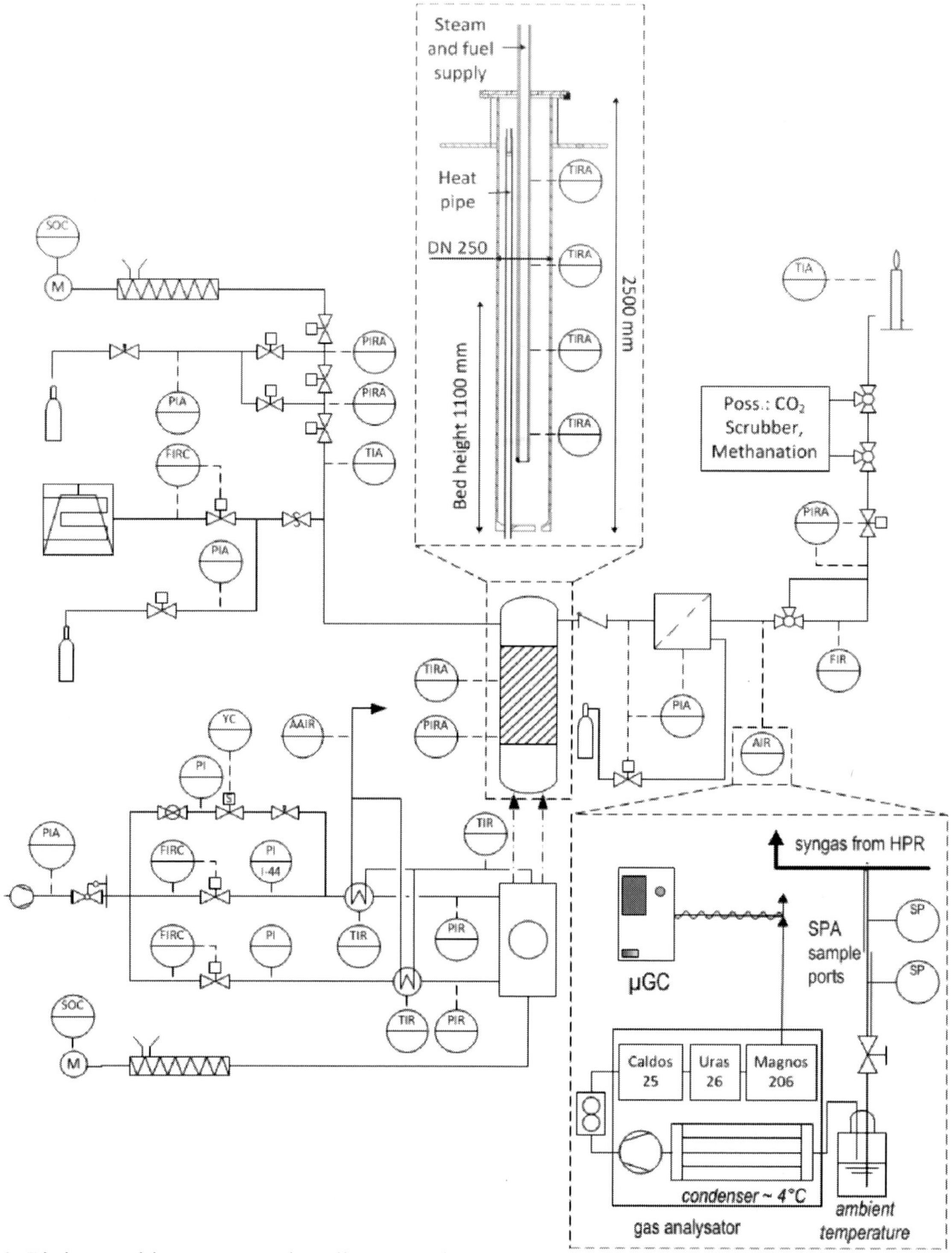

Figure 4: Piping and instrumentation diagram of the Heatpipe Reformer process chain and the gas and tar analysis system

For the combustion, the fuel input is ensured by a screw conveyor. The inlet of primary and secondary air into the combustion chamber is regulated with mass flow controllers fed by compressed air. The lignite supply to the reformer also takes place by a screw conveyor. To pressurize the lignite a lock hopper system with three ball valves is installed above the reformer. The lignite is pressurized by using nitrogen, which also acts as a tracer component for subsequent measurement of the synthesis gas flow. Following the lock hopper system the steam inlet is mounted. The steam supply is guaranteed by a steam generator, it is controlled with a Coriolis flow meter.

After leaving the gasifier the synthesis gas is directed into a candle filter, which removes olivine, ash and char particles. Downstream the candle filter the gas and tar analysis of the synthesis gas takes place, which is described in the following section. Afterwards a pressure-sustaining valve is mounted, which adjusts the reformer pressure. At the end of the process chain the flammable components of the synthesis gas are disposed of in the thermal afterburning.

Results

The measurement campaign in the presented work had a total duration of 72 h with 24 h of gasification operation. After reaching a safe reactor temperature above 800°C the gasification was started. During the measurement campaign five operation points were tested. Table 1 shows the main parameters of the operating points. During operation the fuel input into the gasifier as well as the steam mass flow was kept constant at each operation point while the heat duty of the combustor was adjusted to maintain steady state operation at gasification temperatures between 800 and 840°C.

Table 1: Operation points of the lignite measurement campaign

Name	Pressure [bar]	Heat duty Reformer [kW]	Heat duty furnace [kW]	Steam mass flow [kg s^{-1}]
OP1	1.3	10.8	26.2	4.0
OP2	1.3	15.1	28.9	5.7
OP3	1.5	19.4	34.0	7.3
OP4	4.0	10.8	26.8	4.0
OP4	4.0	15.1	33.2	5.7

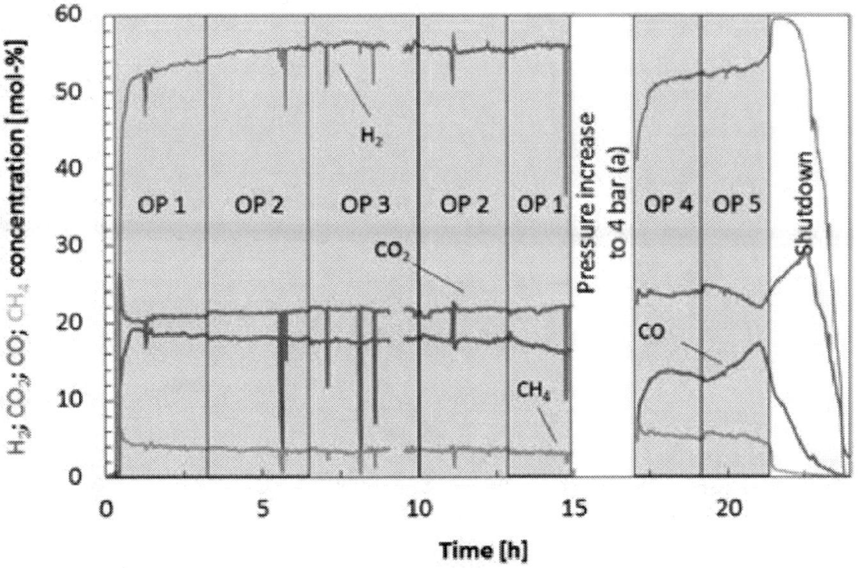

Figure 5: Chronological sequence of the gas composition during the measurement campaign

Figure 5 shows the volumetric concentration of the main gasification products, namely H_2, CO_2, CO and CH_4, as well as chronological sequence of the different operation points. The concentrations remain stable over the measurement campaign after reaching steady state after approx. 5-10 h, which can can be monitored best with the H_2 concentration. Tar content of the synthesis gas was declining with operation temperate of the gasifier and was in the range of 3-5 g/Nm³.

The gasifier showed constant hydrogen contents of 55 % in the synthesis and could safely be operated at 4 bar (a), allowing high hydrogen partial pressures for the membrane separation with nickel membranes. The medium gas composition at OP4 and OP5 was used as a bottle-mixed feed for the following membranes experiments

Membrane experiments

After single membrane experiments for preliminary tests, published in [26] and [27], the gathered information were used for a layout of a membrane bundle for the separation of approx. 1 l/min pure hydrogen. For this purpose 50 Ni 201 tubes with diameter 2.0x0.1 mm from Shaanxi Corp. were used. The following sections cover the measurement techniques for the membrane bundle experiments as well as the analysis of the influence of hydrogen separation on reaction equilibria.

Experimental setup

Figure 6 shows the membrane test rig for the membrane bundle experiments. A mass flow controller was used for the measurement of the hydrogen flow through the membrane bundle. Nitrogen acted as a tracer component to measure the dry volumetric flow rate and as a gas barrier between membrane sealing and measurement cell to prevent condensation at the temperature-sensitive sealing.

To determine the influence of the hydrogen separation on gas composition, the hydrogen permeate flow was fed back to the retentate after the measurement cell and analyzed using a μ-GC "Agilent 490" for the components CO, CO_2, N_2 and CH_4. Although measurement of H_2 is not possible as helium is used as carrier gas in the μ-GC system, it can be calculated as the difference to 100 %.

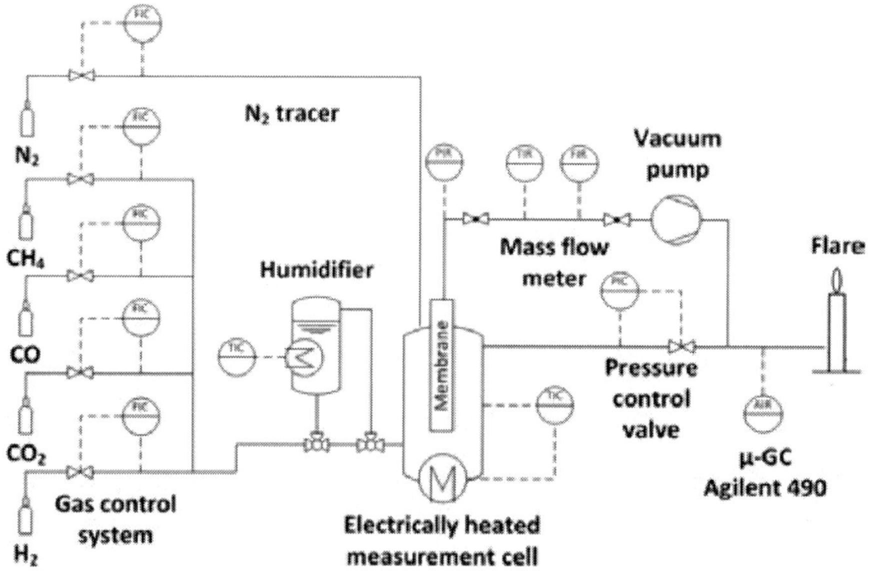

Figure 6: Piping and instrumentation diagram of the test rig for membrane bundle experiments

Results

The gas composition determined in the gasification experiments was used as feed for the measurement cell in the membrane test rig. The composition was 55 % H_2, 24 % CO_2, 16 % CO, 5 % CH_4 with a steam / ratio of 2, which resembles a steam content of 32 %. Figure 7 shows the results from these measurements. The gas composition at a hydrogen separation rate of zero does not resemble the feed composition as the gas mixture reacts due to the high temperatures and the presence of the catalytic nickel surface. Hydrogen separation has a significant influence on the turnover of CH_4 and CO. The CO turnover is negative at low hydrogen separation rates as CH_4 is converted to CO and H_2. As a result, the H_2 and CO_2 content rises due to the ongoing conversion of the feedstock. As proposed in the introduction, it is possible to influence the conversion of the synthesis gas to hydrogen by using nickel membranes.

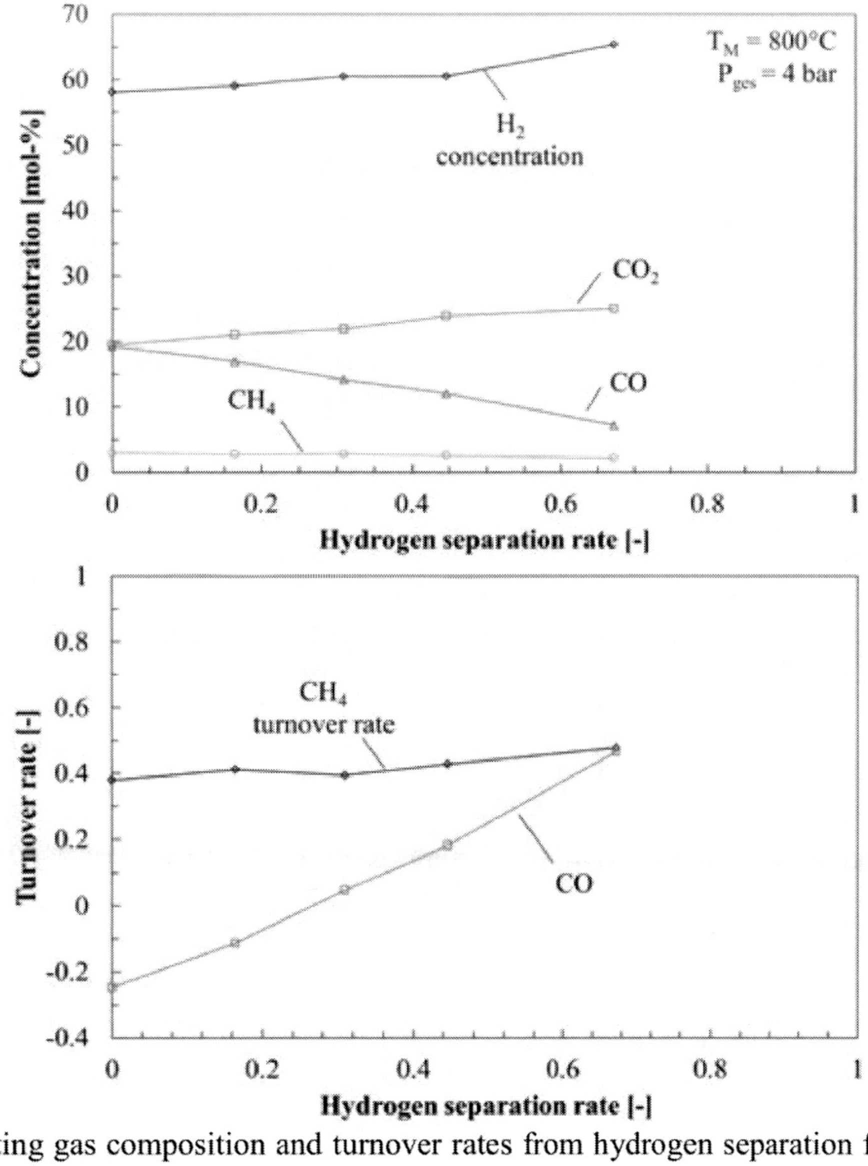

Figure 7: Resulting gas composition and turnover rates from hydrogen separation from bottle-mixed synthesis gas, Membrane temperature: 800°C, pressure 4 bar, gas composition (dry base): 55 % H_2, 24 % CO_2, 16 % CO, 5 % CH_4; steam content 32 %

Conclusion

The paper shows results from the gasification tests with the Heatpipe Reformer at FAU-EVT as well as the results from membrane experiments with a nickel membrane bundle using synthesis gas. As outlined in the state-of-the-art section, the temperature regime of gasification is very challenging for existing membrane systems such as Palladium composite membranes. The usage of nickel membranes was demonstrated and evaluated as an alternative to these systems.

Our experimental results show the good performance of the Heatpipe Reformer system and the suitability of the produced synthesis gas for hydrogen generation. The membrane results show, that the hydrogen yield and the CH_4 and CO turnover can be raised significantly by using nickel membranes.

Ongoing development at FAU-EVT concentrates on the completion and demonstration of a 100 kW SNG process chain with long term experiments. Scientific development is focused on advanced gas cleaning systems, e.g. membranes or scrubbing processes, for a sustainable usage of solid feedstock in energy process engineering

Bibliography

[1] G. Gallmetzer, P. Ackermann, A. Schweiger, T. Kienberger, T. Gröbl, H. Walter, M. Zankl, and M. Kröner, "The agnion Heatpipe-Reformer: operating experiences and evaluation of fuel conversion and syngas composition," *Biomass Convers. Biorefinery*, vol. 2, no. 3, pp. 207–215, May 2012.

[2] S. Karellas, E. Kakaras, T. Papadopoulos, C. Schäfer, and J. Karl, "Hydrogen production from allothermal biomass gasification by means of palladium membranes," *Fuel Process. Technol.*, vol. 89, no. 6, pp. 582–588, Jun. 2008.

[3] C. Schäfer, "Diffusionseigenschaften bestimmter Metalle bei der Hochtemperatur-Wasserstoffabtrennung," Technische Universität München, 2010.

[4] J. M. Leimert, M. Dillig, and J. Karl, "Hydrogen inactivation of liquid metal heat pipes," *Int. J. Heat Mass Transf.*, vol. 92, pp. 920–928, 2016.

[5] P. Quicker, "Dehydrierung von Ethylbenzol zu Styrol," FAU Erlangen-Nuremberg, 2000.

[6] H. Möllenhoff, H. J. Schröter, and H. Jüntgen, "Hydrogen Permeation through Metallic Heat Exchanger Materials during Steam Gasification of Coal," *Zeitschrift für Physikalische Chemie*, vol. 147, no. 1_2. pp. 273–284, 1986.

[7] K. Akamatsu, M. Nakane, T. Sugawara, T. Hattori, and S. Nakao, "Development of a membrane reactor for decomposing hydrogen sulfide into hydrogen using a high-performance amorphous silica membrane," *J. Memb. Sci.*, vol. 325, no. 1, pp. 16–19, Nov. 2008.

[8] Ø. Hatlevik, S. K. Gade, M. K. Keeling, P. M. Thoen, A. P. Davidson, and J. D. Way, "Palladium and palladium alloy membranes for hydrogen separation and production: History, fabrication strategies, and current performance," *Sep. Purif. Technol.*, vol. 73, no. 1, pp. 59–64, May 2010.

[9] S. Yan, H. Maeda, K. Kusakabe, and S. Morooka, "Thin Palladium Membrane Formed in Support Pores by Metal-Organic Chemical Vapor Deposition Method and Application to Hydrogen Separation," *Ind. Eng. Chem. Res.*, vol. 33, pp. 616–622, 1994.

[10] K. Kusakabe, S. Yokoyama, S. Morooka, J.-I. Hayashi, and H. Nagata, "Development of supported thin palladium membrane and application to enhancement of propane aromatixation

on Ga-silicate catalyst," *Chem. Eng. Sci.*, vol. 51, no. 11, pp. 3027–3032, 1996.

[11] E. Fernandez, J. A. Medrano, J. Melendez, M. Parco, J. L. Viviente, M. van Sint Annaland, F. Gallucci, and D. A. Pacheco Tanaka, "Preparation and characterization of metallic supported thin Pd-Ag membranes for hydrogen separation," *Chem. Eng. J.*, 2015.

[12] Y. Guo, H. Zou, H. Wu, L. Zhou, H. Liu, and X. Zhang, "Preparation of palladium membrane by bio-membrane assisted electroless plating for hydrogen separation," *Int. J. Hydrogen Energy*, vol. 39, no. 13, pp. 7069–7076, 2014.

[13] P. P. Mardilovich, Y. She, Y. H. Ma, and M.-H. Rei, "Defect-free palladium membranes on porous stainless-steel support," *AIChE J.*, vol. 44, no. 2, pp. 310–322, Feb. 1998.

[14] G. Xomeritakis and Y.-S. Lin, "CVD synthesis and gas permeation properties of thin palladium/alumina membranes," *AIChE J.*, vol. 44, no. 1, pp. 174–183, 1998.

[15] A. Basile, A. Julianelli, T. Longo, S. Liguori, and M. De Falco, "Pd-based Selective Membrane State-of-the-Art," in *Membrane reactors for hydrogen production processes*, M. De Falco, L. Marrelli, and G. Iaquaniello, Eds. Springer London, 2011, pp. 20–56.

[16] A. Kulprathipanja, G. O. Alptekin, J. L. Falconer, and J. D. Way, "Pd and Pd-Cu membranes: Inhibition of H2 permeation by H 2S," *J. Memb. Sci.*, vol. 254, pp. 49–62, 2005.

[17] H. Amandusson, L. Ekedahl, and H. Dannetun, "The effect of CO and O 2 on hydrogen permeation through a palladium membrane," *Appl. Surf. Sci.*, pp. 259–267, 2000.

[18] C. Chen, "Sulfur tolerance of Pd/Au alloy membranes for hydrogen separation from coal gas," Worcester Polytechnic Institute, 2012.

[19] U. Balachandran, T. H. Lee, C. Y. Park, J. E. Emerson, J. J. Picciolo, and S. E. Dorris, "Dense cermet membranes for hydrogen separation," *Sep. Purif. Technol.*, vol. 121, pp. 54–59, Jan. 2014.

[20] S. Y. Jeon, M. B. Choi, C. N. Park, E. D. Wachsman, and S. J. Song, "High sulfur tolerance dual-functional cermet hydrogen separation membranes," *J. Memb. Sci.*, vol. 382, no. 1–2, pp. 323–327, 2011.

[21] J. M. Polfus, W. Xing, M.-L. Fontaine, C. Denonville, P. P. Henriksen, and R. Bredesen, "Hydrogen separation Membranes based on dense ceramic composites in the La27W5O55.5–LaCrO3 system," *J. Memb. Sci.*, vol. 479, pp. 39–45, Jan. 2015.

[22] V. Gil, J. Gurauskis, C. Kjølseth, K. Wiik, and M.-A. Einarsrud, "Hydrogen permeation in asymmetric La28 − xW4 + xO54 + 3x/2 membranes," *Int. J. Hydrogen Energy*, vol. 38, no. 7, pp. 3087–3091, Mar. 2013.

[23] M. Liu, W. Sun, X. Li, S. Feng, D. Ding, D. Chen, M. Liu, and H. C. Park, "High-performance Ni–BaZr0.1Ce0.7Y0.1Yb0.1O3−δ (BZCYYb) membranes for hydrogen separation," *Int. J. Hydrogen Energy*, vol. 38, pp. 14743–14749, 2013.

[24] Z. Zhu, W. Sun, Y. Dong, Z. Wang, Z. Shi, Q. Zhang, and W. Liu, "Evaluation of hydrogen permeation properties of Ni-Ba(Zr0.7Pr0.1Y0.2)O3−δ cermet membranes," *Int. J. Hydrogen Energy*, vol. 39, no. 22, pp. 11683–11689, Jul. 2014.

[25] J. M. Leimert, P. Treiber, and J. Karl, "The Heatpipe Reformer with optimized combustor design for enhanced cold gas efficiency," *Fuel Process. Technol.*, vol. 141, pp. 68–73, 2016.

[26] J. M. Leimert and J. Karl, "Nickel membranes for in-situ hydrogen separation in high-temperature fluidized bed gasification processes," *Int. J. Hydrogen Energy*, vol. 41, no. 22, pp. 9355–9366, 2016.

[27] J. M. Leimert, A. Bartolf, and J. Karl, "Hydrogen concentration measurement in high temperature applications using hydrogen permeation," *Int. J. Hydrogen Energy*, vol. 41, no. 32, pp. 14360–14367, 2016.

469525 Performance of Transition Metal Catalysts Based on Initial Predictor for Fischer Tropsch Reaction

Abstract:

Performance of transition metal catalysts based on initial predictor for Fischer Tropsch reaction

Authors: Sumegha Godara[1], Suraj Gyawali[1] and Dr. Daniela S. Mainardi[2]

[1]Institute for Micro-manufacturing, Louisiana Tech University

[2]Institute for Micro-manufacturing, Chemical Engineering, Louisiana Tech University

All Fischer-Tropsch mechanisms known to date begin with the adsorption of carbon monoxide followed by its dissociation on a given catalysts surface. Cluster models of Co, Fe, Ni, Pd, Pt, and Ru have been explored using Density Functional Theory. Carbon monoxide adsorption energies were calculated for adsorption on all possible catalytic sites, and the most preferred CO adsorption sites were found in each case. An initial predictor that can be used to anticipate potentially effective catalysts was identified as a percentage difference, using CO adsorption energy results in combination with CO dissociation energies from the previously found sites. A greater catalysts performance is expected when that percentage difference is maximized. On pure clusters, the predictor indicates that ruthenium is expected to be the best catalyst followed by cobalt, in very good agreement with the current knowledge in this field. Hydrogen assisted CO adsorption and formation of hydrocarbons is under study.

Thus, this work presents a mechanism to quickly explore the natural potential of a catalyst material to break the CO bond. The predictor presented here can guide the synthesis of new catalysts, involving modifications of known and currently used catalysts for the Fischer-Tropsch process, and investigations on catalysis in general.

Copyright © American Institute of Chemical Engineers. All rights reserved.

463585 Highly Thermal-Stable and Functional Cellulose Nanocrystals and Nanofibrils Produced Using Fully Recyclable Organic Acids

Abstract:

Here we report the production of highly thermal stable and functional cellulose nanocrystals (CNC) and nanofibrils (CNF) by hydrolysis using concentrated organic acids. Due to their low water solubility, these solid organic acids can be easily recovered after hydrolysis reactions through crystallization at a lower or ambient temperature. When dicarboxylic acids were used, the resultant CNC surface contained carboxylic acid groups which facilitate functionalization and dispersion in aqueous processing. Carboxylic acid group content was 0.1 – 0.4 mmol/g for CNC produced from a bleached eucalyptus kraft pulp (BEP) using oxalic acid at concentrations of 50 – 70 wt%. Onset thermal degradation temperature for the CNC was increased to 322°C from 274°C for the feed BEP fibers, compared with 218°C for CNC produced from the same feed fibers using conventional concentrated sulfuric acid hydrolysis. The low strength (high pKa) of organic acids also resulted in CNC with longer lengths of 200 – 400 nm and higher crystallinity than those produced using mineral acids. Fibrous cellulosic solid residue (FCSR) collected from acid hydrolysis was an excellent feedstock for producing CNF through subsequent mechanical fibrillation with low energy input. The ability to recover the organic acids using a conventional and commercially proven crystallization method makes these organic acids uniquely suited for sustainable and green production of cellulose nanomaterials. The resultant CNC and CNF with high thermal stability and large aspect ratio are excellent for bio-composite applications.

Copyright © American Institute of Chemical Engineers. All rights reserved.

467780 Two-Step Process to Create "Roll-Off" Superamphiphobic Paper Surfaces

Abstract:

The fabrication of anti-wetting surfaces has gained increased attention during recent years; of particular interest is the achievement of superamphiphobic (superhydrophobic and superoleophobic) materials. For anti-wetting applications such as disposable non-stain clothes, low cost medical testing strips or packages for liquids and food, paper has many merits over polymeric or glass containers such as breathability, low cost, recyclability and texture. In this study, the fabrication of paper which repels organic liquids with a surface tension as low as 27.5 mN/m (n-hexadecane) is reported. Paper was exposed to a 30 minute oxygen plasma followed by a 15 minute vapor-phase silanization with fluoroalkyl trichlorosilane. The resulting surfaces cause all test liquids (water, diiodomethane, motor oil and hexadecane) to "bead up" and readily roll off the treated surface. Unlike the widely used approach that relies on the deposition of nanoparticles to achieve superamphiphobic surfaces, this method results in good adhesion between coating and the paper substrate. Anti-wetting behavior and surface stability were determined by contact angle measurements and surface XPS analysis after dynamic impact tests. The surface modification process is simple and does not produce any waste liquid. Furthermore, the superamphiphobic paper has high gas permeability due to pore volume enhancement by plasma etching, but maintains the mechanical flexibility and strength of untreated paper.

Copyright © American Institute of Chemical Engineers. All rights reserved.

467826 3D Printed Nanocellulosic Materials and Their Composite

Abstract:

Cellulose nanocrystals and fibers, which can be extracted from the cell wall of trees, offer a sustainable and greener alternative for petroleum products. Studies have shown nanocellulosic polymer composite materials can exhibit light weight and high barrier properties. These properties have been exploited to form barriers films, reinforcing films, platforms for printable electronics, and many other cellulosic products. Nevertheless, it is still a challenge to find compatible nanocellulose-curable resins composite, and design a facile and scalable approach for fabricating composite materials into functional structures. In this work, we introduced cellulose nanocrystals into an acrylate based photo-curable resin, then projection stereolithography was utilized to 3D print nanocellulose composite structures. A custom built direct ink write 3D printer was also utilized to 3D print cellulose nanocrystals and nanofiber, which resulted in pure cellulosic 3D aerogels after freeze dry processing. The mechanical properties of the resultant 3D printed structures were characterized. This work established the printing platform for generating nanocellulosic 3D structure and aerogel, which enabled future possibility for incorporating targeted functionalities into these structures.

Copyright © American Institute of Chemical Engineers. All rights reserved.

466469 Modifying the Surface Properties of Wood Using ATRP Grafting Polymerization

Abstract:

Wood is a well-known bio-based material which can be used for a wide range of applications, with a major focus on its utilization as a building material. In order to profit from its natural properties and to develop smart and innovative materials, a variety of chemical modifications can be used. Cabane et al. have shown that wood can be modified by in-situ grafting of synthetic polymers enhancing the natural properties of wood and making it competitive with other available materials [1]. We recently extended the modification of wood through the grafting of fluorinated acrylates by surface-initiated ATRP (Fig 1A). ATRP is a living polymerization technique allowing for a better control over grafted polymer chain length. The polymer chains are grafted from the wood biopolymers through a two-step reaction. In the first step the ATRP initiator is grafted onto the wood surface (via esterification), and it is used in the second step to grow the polymer chains.

Several techniques were used to characterize the wood properties after the polymerization. Amongst others, the most outstanding results obtained were in regard to the wettability. Through the functionalization of wood with a fluorinated polymer, and following the examples of superhydrophobic surfaces found in nature, we have acquired a renewable material that is both superhydrophobic and oleophilic. The contact angles of water on the wood surface measured before and after modification increase drastically with values raising from 73° to 180 ° (Fig 1B). This increase is both due to the chemical modification and the surface roughness acquired by the combination of the both the natural roughness of wood and the grafted polymer chains.

Moreover, quaternary amine acrylates were grafted onto the wood in order to further study the wettability given by the combination of the natural structure of wood and the chemical and physical properties of the polymer introduced. Both quaternary amines and fluorinated acrylates have been separately and successfully grafted into wood (Fig. 1C) offering

different wettabilities (Fig 1D). In further work we would like to extend the methods used to copolymerize fluorinated and quaternary amine acrylates, combining the properties of both polymers. By this copolymerization it would be possible to obtain a material that is both hydrophilic and oleophobic [2]. This new material could be used to extract the oil present in contaminated water resources.

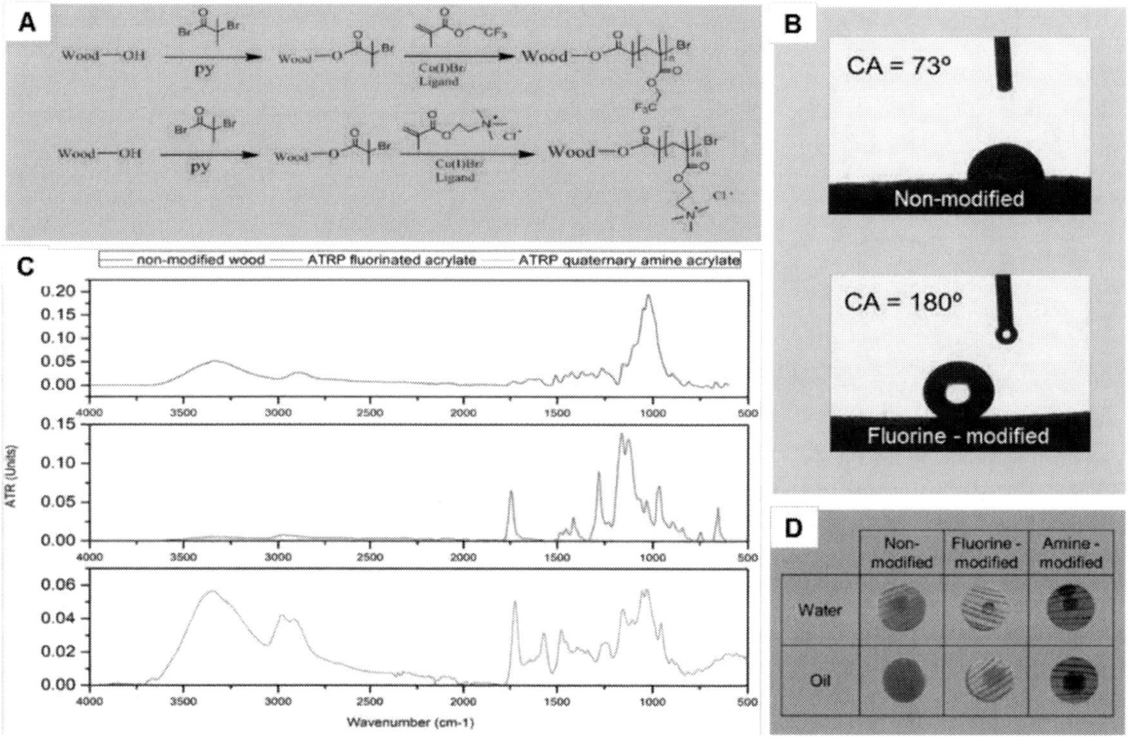

Fig. 1: A) Wood modification scheme; B) Contact angle measurements of water on the wood surface; C) FTIR spectra; D) Wettability of wood with various liquids.

[1]. E. Cabane, T. Keplinger, V. Merk, P. Hass, I. Burgert, *ChemSusChem* **2014**, 7, 1020-1025.

[2]. P. Brown, O. Atkinson, J. Badyal, *ACS Applied Materials & Interfaces* **2014**, 6, 7504-7511.

Copyright © American Institute of Chemical Engineers. All rights reserved.

458027 Synthesis of Cellulose-Based Injectable Hydrogel Composite for pH-Responsive Drug Delivery

Abstract:

Abstract: Localized drug delivery, and prolonged, stimulus-driven release are approaches to maximize the function of drugs and minimize their side effects. This study developed a novel cellulose-based injectable hydrogel composite system containing pH responsive poly (ethylene oxide)-block-poly (2-(diisopropylamino) ethyl methacrylate) (PEO-b-PDPA) copolymer micelles for localized delivery and pH-triggered release. First, covalently in situ generated hydrogels were constructed by mixing hydrazide-modified carboxymethyl cellulose (CMC-NH_2) with oxidized carboxymethyl cellulose (CMC-CHO). The resultant hydrogels could be well tuned in the aspects of gelation time, degradation profile, and mechanical properties. Then, the pH-responsible copolymer PEO-b-PDPA was introduced to enhance the loading of hydrophobic substances in hydrogels as well as a pH-triggered release, which was demonstrated by fluorescent spectroscopy using Nile Red as a model drug. In addition, the results demonstrated that the hydrogel itself could act as a second diffusion barrier to reduce the drug release rate. The cell viability tests showed that both CMC-NH2 and hydrogels possessed excellent biocompatibility, while CMC-CHO was cytotoxic at a relative high concentration.

Copyright © American Institute of Chemical Engineers. All rights reserved.

464660 Preparation of Whole Biomass Aerogels from Douglas Fir Using Molten Salt Hydrate As Solvent

Abstract:

Aerogel is a material with porous network structure, low density, high specific surface area, and low thermal conductivity. Aerogel has been used for thermal insulator, supercapacitor, storing medium, filter material for extremely fine particles, drug delivery systems, and drug coating. In the present study, whole biomass aerogels with high porosity, large surface area and low-density open pore 3D structure were prepared from Douglas fir without hemicellulose removal and delignification process. Lithium bromide molten hydrate was used to dissolve â€“ regenerate the cellulose or biomass to form cellulosic network structure. The results indicated that the density of the aerogels prepared from Douglas fir whole biomass ranged from 4.6 to 29 mg/cm^3 with porosity from 99.7% to 98.1%. The aerogel properties changes with biomass pretreatment, different reaction conditions, and biomass loading were investigated.

Copyright © American Institute of Chemical Engineers. All rights reserved.

452776 Production of Light Olefins from Protein-Rich Microalgae By Hydrothermal Liquefaction and Sequential Catalytic Cracking

Abstract:

In recent years, biomass resources have gained increasing attention as sustainable feedstock for production of energy fuels and chemicals [1]. Algae is a very promising source as the third-generation biomass, following after the first-generation biomass (edible crops, sugars, and starches) and the second-generation biomass (inedible lignocellulose). Algal biomass can be converted into biofuels such as biogas, bioethanol, biodiesel, and bio-oils by anaerobic digestion, fermentation, gasification, liquefaction, pyrolysis, and so on [2]. Although hydrothermal liquefaction (HTL) is considered the most promising method for conversion of wet microalgae to often called â€˜bio-oilâ€™, industrial viability of microalgae bio-oil production still has a lot of challenges to overcome [3]. In this study, we propose a unique method to produce light olefins such as ethylene, propylene, and butene from protein-rich microalgae.

Dry microalgae cells of Spirulina sp. was purchased from DIC LIFETEC Co., Ltd. (Tokyo, Japan). Spirulina sp. consisted of 4.7 wt% moisture, 6.1 wt% ash, 12.6 wt% carbohydrates, 7.8 wt% lipids, and 68.8 wt% protein. It contained 53.0 wt% C, 8.0 wt% H, 11.5 wt% N, 0.9 wt% S, and 26.6 wt% O (dry ash free). All other materials were purchased from commercial suppliers and used as received without any pretreatment or purification. Hydrothermal reaction experiments were carried out in a 100-mL hastelloy autoclave to obtain both water soluble and insoluble fractions. A fixed bed reaction system was used for catalytic cracking of water soluble fraction. The products were analysed using a gas chromatograph, high-performance liquid chromatographs, a total organic carbon analyser, a total nitrogen analyser, an elemental analyser, and etc. The yields of the products (mol%/C) were calculated based on the carbon content.

We demonstrated the hydrothermal liquefaction of *Spirulina sp.* at from 250 °C to 300 °C and sequential catalytic cracking using a HZSM-5 zeolite catalyst at 600 °C. We show that we obtained light olefins as major products along

with aromatics and light alkanes as minor products in the presentation. The possible conversion pathways of carbohydrates, lipid, and protein to bio-oils in our system are also proposed. This research provides a useful production method of light olefins from protein-rich microalgae.

References

[1] a) J. N. Chheda, G. W. Huber and J. A. Dumesic, Angewandte Chemie-International Edition 2007, 46, 7164-7183; b) A. J. Ragauskas, C. K. Williams, B. H. Davison, G. Britovsek, J. Cairney, C. A. Eckert, W. J. Frederick, J. P. Hallett, D. J. Leak, C. L. Liotta, J. R. Mielenz, R. Murphy, R. Templer and T. Tschaplinski, Science 2006, 311, 484-489.

[2] L. Brennan and P. Owende, Renewable & Sustainable Energy Reviews 2010, 14, 557-577.

[3] a) D. López Barreiro, W. Prins, F. Ronsse and W. Brilman, Biomass and Bioenergy 2013, 53, 113-127; b) Y. Guo, T. Yeh, W. Song, D. Xu and S. Wang, Renewable and Sustainable Energy Reviews 2015, 48, 776-790.

Copyright © American Institute of Chemical Engineers. All rights reserved.

454715 Selective Adsorption of Au (III) By Epichlorohydrin/Thiourea Modified Porous Alginate Beads

Abstract:

Selective adsorption of Au (III) by epichlorohydrin/thiourea modified porous alginate beads

Xiangpeng Gao,[a] Yan Zhang,[a*] Yuming Zhao[b]

[a] Department of Process Engineering, Memorial University of Newfoundland

[b] Department of Chemistry, Memorial University of Newfoundland

Abstract

Sodium alginate is an anionic polysaccharide distributed widely in the cell walls of brown algae, which can be extracted by treating algae with sodium carbonate solution and recovered by precipitation. Sodium alginate and its derivatives can be used as biosorbents to retain toxic metal ions, or recover precious metal ions from industrial wastewater. However, the dissolution of sodium alginate in aqueous solution has limited its application as a sorbent. Moreover, powder form of sodium alginate is not suitable to be packed in a column for applications at preparative scale because they may cause clogging problems, a huge pressure drop or serious hydrodynamic limitations. Therefore beads are regarded as the best form of biopolymer based sorbents for metal uptake from aqueous solutions. So far, most of the studies in biosorption area focused on the sorption of monometallic system. A major complication in the application of biosorption to industrial conditions is the presence of other metal ions in the wastewater to be treated. The effect of the co-existing metal ions on the selective adsorption of the meal of interest may be different under different scenarios. In this study, porous alginate beads were prepared by modifying epichlorohydrin and thiourea with sodium alginate, and then dropped into calcium chloride solution to form beads. The epichlorohydrin/thiourea modified alginate beads were soaked in tetraethyl orthosilicate (TEOS) sol and finally freeze dried to generate pores. Selective adsorption of Au (III) from mixed metal solutions under different acidic conditions was studied. Mathematical models were used to describe the adsorption isotherm and kinetics of Au(III). FT-IR, SEM-EDS, and XPS analyses were used to characterize the sorbent and identify the functional groups and potential biosorption mechanisms that contributed to the selective adsorption of Au (III).

Keywords: Selective adsorption; Alginate; Porous biosrobent

Copyright © American Institute of Chemical Engineers. All rights reserved.

461820 Hydrothermal Carbonization of FOOD Wastes from LOCAL Eatery

Abstract:

Since 2014, it has been estimated that the world has produced nearly 2 billion tons of garbage also known as municipal solid waste (MSW) and it is further projected that this generated waste will be increased by nearly five times by 2050 due to large economic and urban growth of developing countries. In the U.S, a typical garbage disposal fee of $20-50 per month, varies from county-to-county, applies to every household. Garbage contains mainly food waste and the composition can add up to 70 %. Except from some cases, where food waste garbage has been composted or used as animal feed, it ends up in a landfill, where it contributes to environmental pollution via greenhouse gases (mainly methane and carbon dioxide). Large metropolitan cities like Los Angeles, New York etc. are facing serious waste disposal challenges with the lack of adequate lands. Although garbage has huge potential for energy production, converting garbage to energy in both efficiently and economically is yet to be discovered.

Hydrothermal carbonization (HTC) is an exceptional treatment process for biofuel production, as residual water is the reaction media for HTC. Therefore, the wet biomass need not be dried prior to the HTC. Subcritical water at around 200-260 °C has the highest ionic constant, which means it is quite reactive. As a result, within 5 minutes of reaction with biomass, solid-hydrophobic biofuel is formed along with liquid and gas byproducts. The solid product, also called hydrochar, is quite hydrophobic, friable, and more importantly energy dense; an increase in energy density means that less mass is required to produce the same amount of energy. Pelletization of the hydrochar allows for more energy densification as well as better storage of the fuel. Moreover, the exothermic HTC reaction ensures the minimum energy input for biofuel production. Researchers around the world have been developing HTC over the last decade for woody biomass and agricultural residues.

The main goal of this study is to apply HTC to local eatery wastes for producing clean and sustainable energy. Waste disposal issues will be automatically resolved with this proposed process. In this study, a 2L batch Parr reactor was used for HTC for recycle, compostable, and trash bins wastes at 200, 230, and 260 °C at their water saturation pressure. Solid hydrochar, HTC process liquid, and gaseous products were analyzed for their physicochemical properties. Based on batch HTC results, a simple techno-economic analysis was performed using ChemCAD 6.0 simulation. The simulation results indicate that higher HTC temperature is favorable for solid fuel production, however, such high-pressure and high-temperature system requires high capital investment.

Copyright © American Institute of Chemical Engineers. All rights reserved.

Production of Jet Fuel and Butadiene from Lignocellulose and Lignocellulosic Platform Chemicals

Cui Xingkai, Zhao Xuebing*, Liu Dehua

Institute of Applied Chemistry, Department of Chemical engineering, Tsinghua University

Correspondence: zhaoxb@mail.tsinghua.edu.cn

Introduction

With the reduction of fossil energy reserves, increasing concerns of environmental pollutions and greenhouse gases emissions, more and more attentions have been payed to developing green and sustainable chemicals and fuels. Lignocellulose, which is generated numerously every year in nature, is a feasible alternative to fossil resources. Three strategies are usually applied to produce fuels from lignocellulose: (i) Pyrolysis to produce bio-oil and fuels are produced by bio-oil hydrogenation processing (ii) Gasification to produce syngas and then F-T Synthesis to get fuels; (iii) Pretreatment, hydrolysis and dehydration to generate platform chemicals, which can be converted to liquid fuels by bio-processing or chemical processing [1].

A process employing solid acid catalyst preparation, dehydration, C-C coupling techniques and hydrodeoxygenation (HDO) is being developed to convert lignocellulosic platform chemicals (including furfural, hydroxymethylfurfural (HMF) and 2,3-butanediol) to methyl ethyl ketone (MEK), butadiene and jet fuel. First, lignocellulosic materials, such as bagasse and straw, are converted to solid acid by Incomplete carbonization and H_2SO_4 sulfonation subsequently [2]. Next, the prepared lignocellulose based catalyst is used to catalyze 2,3-butanediol dehydration reaction that can produce methyl ethyl ketone (MEK) at low temperatures and butadiene at high temperatures. Further, by undergoing aldol condensation[3] and hydrodeoxygenation (HDO) subsequently, MEK and furfural (or HMF) can be converted to C9-C14 (or C10-C16) linear and branched hydrocarbons that meet the requirements of jet fuels. The aldol condensation kinetics are also investigated to make molecular size distributions in the final hydrocarbons controllable.

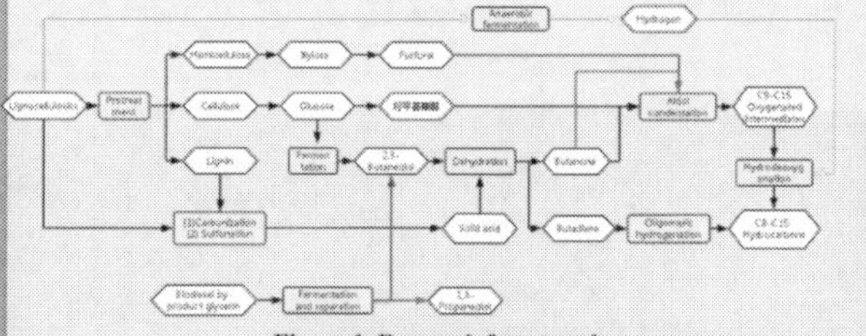

Figure 1. Research framework

Results and Discussions

Preparation of lignocellulosic-based solid acid

Sugarcane bagasse and lignocellulosic residues after pretreatment, simultaneous saccharification and fermentationis (SSF) are used as feedstocks for solid acid preparation. Firstly, the feedstocks are incompletely carbonized with protection of nitrogen. Secondly, materials from the first step are sulfonated by concentrated sulfuric acid.

Table 1. Changes in the composition at different carbonization temperatures

(Carbonization time: 0.5h, nitrogen flow rate: 200mL / min)

Temperature(℃)	285	315	345	375	405	435	465
Mass before carbonization(g)	10.72	12.54	11.28	10.23	11.46	11.77	12.12
Mass after carbonization(g)	7.91	7.14	4.45	3.19	3.21	3.1	2.9
The percentage reduction in mass(%)	26.21	43.06	60.55	68.82	71.99	73.66	76.07
Cellulose content (w/w%)	38.8	47	36.5	1.38	1.29	1.09	1.23
Hemicellulose content(w/w%)	27.5	10.8	3.14	0	0	0	0

- At lower temperatures, hemicellulose removal is dominant
- When the temperature is higher than 345 ℃, almost all the cellulose and hemicellulose is removed.

Dehydration of 2,3-butanediol to produce methyl ethyl ketone

Dehydration reaction is carried out in a round-bottomed flask with magnetic stirring, methyl ethyl ketone is separated online with a condenser. Bagasse-based solid acid is used as catalyst. Results are displayed in table 2.

In order to to increase yield of methyl ethyl ketone, lower carbonization temperature, longer carbonization time and shorter sulfonation time should be adopted. For bagasse raw material, optimal conditions for preparing the solid acid is: carbonization temperature 345 ℃, carbonization time 1.5h and Sulfonation time 20min. In addition, the catalyst exhibit good reuse stability. 82% activity remains after 4 reuses.

Table 2. Different conditions for preparing solid acid and comparison of their catalytic performance

(Reaction conditions: 2.5g catalyst, 50g 2,3-butanediol, reaction temperature is 168 ℃, reaction time 6h, stirring speed 300rpm)

Carbonization temperature(℃)	Carbonization time(h)	Sulfonation time(h)	Conversion of 2,3-butanediol(%)	Selectivity of MEK(%)	Yield of MEK (%)
315	0.5	8	61.97	50.89	31.53
345	0.5	8	63.82	73.37	46.83
375	0.5	8	67.81	60.17	40.80
405	0.5	8	69.33	54.00	37.44
435	0.5	8	66.29	53.76	33.98
345	0.17	8	56.60	77.61	43.94
345	1	8	63.06	73.42	46.30
345	1.5	8	61.78	81.82	50.56
345	1.5	0.33	68.55	76.47	52.42
345	1.5	0.5	60.44	81.83	49.46
345	1.5	1	55.24	86.12	47.57
345	1.5	3	56.34	84.10	47.38

Aldol condensation of methyl ethyl ketone and furfural

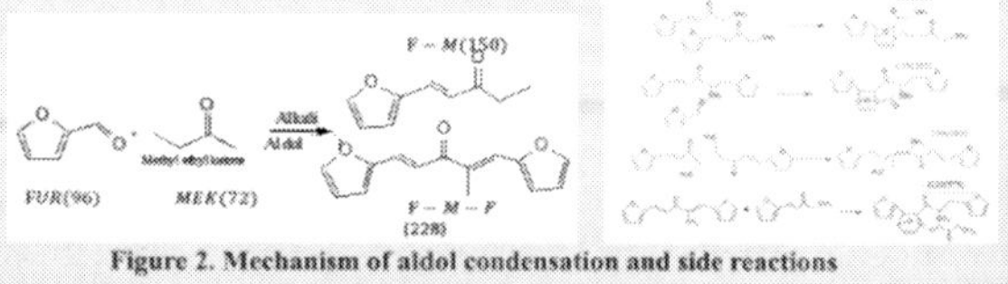

Figure 2. Mechanism of aldol condensation and side reactions

Aldol condensation is carried as two-phase reaction, the organic phase take tetrahydrofuran as solvent, furfural and methyl ethyl ketone are dissolved; the aqueous phase is aqueous sodium hydroxide solution that serves as catalyst; reaction is carried out in a shaker at 300 rpm, the reaction time was 24h, samples taken are analyzed by gas chromatography.

Table 3. Aldol condensation results of furfural and methyl ethyl ketone

Mass ratio of organic to aqueous phase	Furfural in organic phase (w %)	Molar ratio of furfural to MEK	Molar ratio of NaOH to furfural	Reaction temperature (℃)	8h				24h			
					Conversion of furfural (%)	Selectivity of F-M(%)	Selectivity of F-M-F(%)	Selectivity of heavy condensates (%)	Conversion of furfural (%)	Selectivity of F-M(%)	Selectivity of F-M-F(%)	Selectivity of heavy condensates (%)
1:1	13	2:1	2.3	25	65.64	75.36	21.21	3.43	79.00	47.93	40.45	11.61
1:1	13	1:1	2.3	25	89.02	86.91	10.07	3.03	96.09	78.80	13.52	7.68
1:1	13	2:5	2.3	25	94.16	98.01	1.99	0	99.43	92.0	3.60	4.41
1:1	13	1:1	1	25	96.48	85.06	12.48	2.46	99.64	81.81	14.12	4.06
1:1	20	1:1	2.3	25	89.31	80.90	9.82	9.28	96.44	66.01	12.67	21.32
2:1	13	1:1	1	25	84.31	88.49	7.03	4.47	88.97	78.75	9.22	12.03
1:1	13	1:1	1	40	99.27	84.91	12.37	2.71	100	82.28	12.04	5.68
1:1	13	1:1	1	55	100	79.75	15.35	4.90	100	75.47	14.47	10.06

To improve the conversion of furfural and selectivity of F-M and F-M-F, and control selectivity of heavy condensates, following strategies need to be adopted: lower mass ratio of organic to aqueous phase, lower furfural fraction in organic phase, lower molar ratio of furfural to MEK and lower molar ratio of NaOH to furfural.

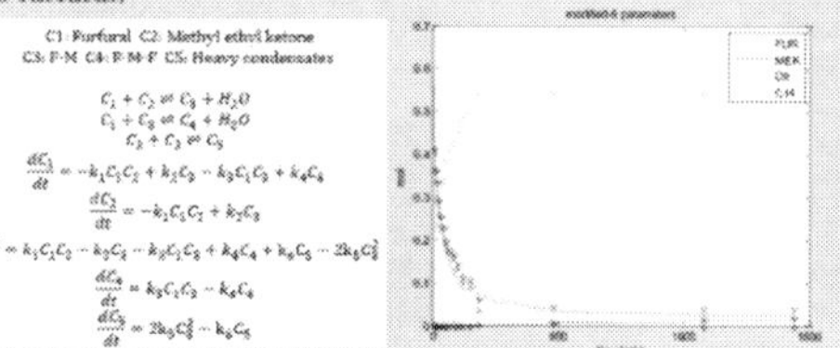

Figure 3. Kinetic modelling of aldol condensation

(simulation work is based on reaction conditions in the 5th row in table 3)

Conclusions

- Bagasse-based solid acid exhibit good catalytic activity and reuse stability in converting 2,3-butanediol to methyl ethyl ketone
- With catalysis of NaOH solution, aldol condensation of furfural and methyl ethyl ketone gives high furfural conversion, and distribution carbon chain length in condensates can be adjusted by reaction conditions.

References

1. John R. Regalbuto. Cellulosic Biofuels--Got Gasoline? *Science*, 2009, 325, 822.
2. Wenyong Lou et al. A highly active bagasse-derived solid acid catalyst with properties suitable for production of biodiesel. ChemSusChem 2012, 5, 1533-1541.
3. Rong Xing et al. Production of jet and diesel fuel range alkanes from waste hemicellulose-derived aqueous solutions. *Green Chem.*, 2010, 12, 1933–1946.

465838 Macroalgae Pretreatment Using Peg-Linked Dicationic Acidic Ionic Liquids

Abstract:

Macroalgae has become a sustainable source of carbohydrates for the production of bio-fuel and bio-based chemicals. Compared with lignocellulosic biomass, macroalgae has higher photosynthetic efficiency, higher yields, lower lignin content and requires a simpler environment for its cultivation. Saccharification of macroalgae is commonly performed in a pressurized system in the presence of an acid catalyst. But ionic liquids (ILs) have been gaining popularity as effective and environmentally benign solvents for biomass as it enables hydrolysis to be performed at atmospheric conditions.

Task specific ILs are currently being explored in biomass processing owing to their versatility and multi-functionality (Olivier-Bourbigou et al., 2010). Particularly, Bronsted acidic ILs could replace traditional mineral acids such as sulfuric and hydrochloric acids. Typically, mono-cationic ILs are used to initially dissolve the biomass forming a homogeneous solution which is subsequently hydrolyzed using either enzymes or an acid catalyst. Meanwhile, dicationic ILs have various advantages over mono-cationic ILs such as higher thermal stability, decreased toxicity and multi-functionality. Particularly, ether-functionalized ILs tend to exhibit lower viscosities than their aliphatic counterparts. Moreover, the oxygen atoms embedded in the ether chains could act as hydrogen bond acceptors promoting biomass dissolution (Zhao et al., 2008). Dicationic ILs with acidic counter-anions are scarcely reported in biomass processing.

In this work, dicationic acidic ILs (DAILs) with varying poly-ethylene glycol (PEG) linker chains were synthesized, characterized and utilized for the conversion of *Gelidium amansii*, a red macroalgae into its corresponding monomeric sugars. The structures of PEG-DAILs having an acidic counter-anion (i.e. $[2HSO_4]^-$) were confirmed using

spectroscopic techniques. Results showed the liquid form of PEG-DAILs, which are thermally stable up to T_{peak} < 350 °C. PEG-DAILs with short ether chains (e.g. Di-, Tri- and Tetra-ethylene glycol) exhibited faster reaction rates than those with long ether chains (e.g. PEG400 and PEG600). Parameters such as particle size, biomass loading, reaction temperature and time were varied to achieve optimal sugar yields. Results showed $[Tri\text{-}EG\text{-}(MIm)_2]\ 2HSO_4$ as the best PEG-DAIL for maximum sugar release from *G. amansii*. Overall findings show the potential of PEG-DAILs for one-pot dissolution and hydrolysis of macroalgae to sugars.This work was supported by the National Research Foundation of Korea (NRF) grant funded by the Ministry of Education (No. 2009-0093816) and by the Ministry of Science, ICT & Future Planning (No. 2015R1C1A2A01054605).

References:

Olivier-Bourbigou, H., Magna, L., Morvan, D., *Appl. Catal. A-Gen.*, 373 (2010) 1.

Zhao, H., Baker, G. A., Song, Z., Olubajo, O., Crittle, T., Peters, D., *Green Chem.*, 10 (2008) 696.

Copyright © American Institute of Chemical Engineers. All rights reserved.

467504 Fast Pyrolysis Bio-Oil As Precursor of Thermosetting Resins

Abstract:

The use of biomass-based resources is currently a hot topic for both academic and industrial research as an alternative to mitigate the strong dependence on fossil carbon in the chemical and petrochemical industries. The U.S. Department of Energy (DOE) and the U.S. Department of Agriculture (USDA) have prioritized the development of bioenergy and bio-products; and they have the goal to produce 18% of the current U.S. chemical commodities from biomass by 2020, and 25% by 2030. The macromolecular chemistry based on lingocellulosic feedstock represents one of the answers to the quest for polymeric materials capable of replacing their fossil-based counterparts. In particular, fast pyrolysis bio oil is an excellent candidate for the bio-sourcing of polymeric resin. Their competitive cost, worldwide availability and built-in functionality have catapulted it use as a source of macromonomers for polymer applications. The development of this area has been carried out mainly through the reaction and functionalization of hydroxyl groups in the bio oil to produce thermosetting resins such as: epoxy, phenolic, polyurethanes, etc. Thus, the main purpose of this project is to cover the major aspects related to the chemical synthesis, physical-chemical characterization and study of thermo-mechanical properties of bio-based resin, where highly functionalized components are synthetized by chemical modification. Results showed that this new bio-based polymeric systems display interesting properties that are close to their commercial counterparts.

Copyright © American Institute of Chemical Engineers. All rights reserved.

463504 Micro/Nano Lignocellulosic Fibrils (MNLCF) Aerogels from Coconut and Oil Palm Tree Residuals and Application for Environmental Remediation

Abstract:

The extensive utilization of oil palm and coir plantations worldwide has led to a substantial generation of residual biomass[1]. Therefore, an integrated utilization of related biomass streams is a necessity that, in fact, can become a source of valuable natural fibers. In addition, replanting operations and environmental issues can be avoided this way. The lignocellulose sources used in this study, include coir fibers and oil palm empty fruit bunches (EFB), both of which have high microfibril angle ($\sim 45^0$) that imparts high breaking strain and low elastic modulus, making them useful for producing ropes, mats and fishing nets[2,3]. Coir and EFB fibers are cheap and a renewable resource but a vast supply is wasted due to a lack of better utilization. These fibers have high lignin content, approximately 43% and 36% dry weight in coir and EFB, respectively. Lignin is an undesirable component in paper manufacturing as it causes yellowing of the paper and hence, coir and EFB fibers do not find their use in high-volume papermaking. In fact, the current research on nanocellulose also focuses on producing lignin-free cellulose nanofibers which leaves a high carbon footprint on the environment[4,5]. Currently, lignin is utilized for their high calorific value which is a result of a cross-linked phenolic structure[6]. Recently, lignin has been shown to act as cementing material between cellulose nanofibers[7]. In addition, it has also been demonstrated that the porosity, hydrophilicity and the barrier properties of nanopapers or films of nanocellulose can be tuned via lignin content.

A high microfibril angle and a high lignin content, limits the applications of coir and EFB fibers. In the present study, we approach this problem by isolating micro/nano fibrils from the source fiber, thus foregoing effects related to the microfibril angle. We also target retaining the native lignin of coir and EFB by subjecting them to mild pulping conditions via hydrothermal treatment (see Table 1). A recently reported micro-emulsion method is used to facilitate defibrillation of the fibers[8]. The micro-emulsions swell and delaminate different structures of the fibers by disrupting intra-molecular hydrogen bonding. The micro/nano lignocellulosic fibrils (MNLCF) are obtained through microfluidization. A detailed characterization of the fiber defibrillation process is carried out via composition analysis, scanning electron microscopy (SEM), atomic force microscopy (AFM) and X-ray diffraction (XRD). The migration of lignin from cell wall to the defibrillated fibers is studied via confocal microscopy. The thermal properties of MNLCF is studied via thermogravimetric analyses (TGA) and differential scanning calorimetry (DSC). Our detailed analyses provide useful information to identify potential applications for the produced MNLCF. As a proof of concept, water-resistant and durable aerogels are synthesized from MNLCF via freeze drying (see Figure 1) and tested for adsorption of Zn(II), Pb(II) and Cd(II) salts.

Table 1: Chemical composition of fiber precursor before and after the treatment. All values are in %

Sample	Extractives	Lignin	Cellulose	Hemicellulose
Coir (raw material)	0.5	43	30	23
Coir (Hydrothermal treatment)	<0.1	34	56	7
Coir (microemulsion treatment)	<0.1	31	56	5
EFB (raw material)	0.8	36	34	20
EFB (Hydrothermal treatment)	<0.5	29	58	4
EFB (microemulsion treatment)	<0.5	24	67	3

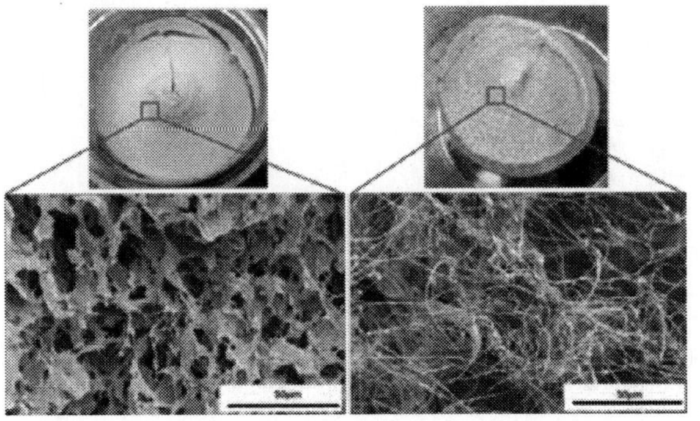

Figure 1: As-prepared aerogels from EFB (left) and coir (right) micro/nano lignocellulosic fibrils (MNLCF) along with their SEM images. The scale bar corresponds to 50μm

References:

1. Ferrer A, Vega A, Rodríguez A, Jiménez L. Acetosolv pulping for the fractionation of empty fruit bunches from palm oil industry. *Bioresour Technol*. 2013;132(November 2015):115-120. doi:10.1016/j.biortech.2012.12.189.

2. Khalil HPSA, Alwani MS, Omar AKM. Chemical composition, anatomy, lignin distribution, and cell wall structure of Malaysian plant waste fibers. *BioResources*. 2006;1(2):220-232. doi:10.15376/biores.1.2.220-232.

3. Satyanarayana KG, Kulkarni AG, Rohatgi PK. Structure and Properties of Coir Fibres. *Proc Indian Acad Sci Ser Eng Sci*. 1981;V 4 PT4(February):419-436. doi:10.1007/BF02896344.

4. Abraham E, Deepa B, Pothan L a., et al. Extraction of nanocellulose fibrils from lignocellulosic fibres: A novel approach. *Carbohydr Polym*. 2011;86(4):1468-1475. doi:10.1016/j.carbpol.2011.06.034.

5. Deepa B, Abraham E, Cordeiro N, et al. Utilization of various lignocellulosic biomass for the production of nanocellulose: a comparative study. *Cellulose*. 2015;22(2):1075-1090. doi:10.1007/s10570-015-0554-x.

6. Vanholme R, Demedts B, Morreel K, Ralph J, Boerjan W. Lignin Biosynthesis and Structure. *Plant Physiol*. 2010;153(3):895-905. doi:10.1104/pp.110.155119.

7. Rojo E, Peresin MS, Sampson WW, et al. Comprehensive elucidation of the effect of residual lignin on the physical, barrier, mechanical and surface properties of nanocellulose film. *Green*. 2015:1853-1866. doi:10.1039/c4gc02398f.

8. Carrillo CA, Laine J, Rojas OJ. Microemulsion Systems for Fiber Deconstruction into Cellulose Nano fi brils. *Appl Mater Interfaces*. 2014;6:22622-22627. doi:10.1021/am5067332 |.

Copyright © American Institute of Chemical Engineers. All rights reserved.

466246 *Tuning Ionic Liquids for Low Cost: Applications in Lignocellulose Deconstruction*

Abstract:

Ionic liquids (ILs) have proven to be highly tunable 'designer solvents' capable of a wide range of exciting chemistries. However, industrial application at large scale is hampered by high solvent cost. This cost is, however, a tunable feature of the solvent itself – provided the ion selection is handled with a careful eye aimed at limiting synthetic complexity. Lowering the solvent cost will increase the attractive opportunities of ILs for bulk processing of lower cost end products – including such applications as biofuels. One of the key challenges in biorefining is the initial separation or deconstruction of lignocellulosic feedstock into separate components. ILs offer unique advantages in this area, due to their unusual thermochemical properties. However, there are serious concerns about the economic viability of their use due to the very high cost of most ionic liquids (> $50/kg). We have overcome this by redesigning the IL based deconstruction process to use low-cost, acidic ILs for lignin dissolution rather than cellulose dissolution, yielding filterable cellulose and a dissolved lignin for precipitation or conversion to high-value chemicals. We have found that processability of the cellulose is high and lignin recoveries near quantitative.

We use a range of 'protic' ILs, the family typically used in IL industrial processes, because their simple acid-base chemistry results in a simple and cheap synthesis, with a cost (< $1/kg) similar to common organic solvents such as acetone or toluene. This presentation will discuss how ionic liquids can be 'tuned' to control cost structure of the final solvent, and what implications this will have for the chemical processes involved. The impact of the solvent on large-scale applications, such as biomass pretreatment, will be discussed, with a focus on performance and process considerations such as how the ILs maintain solvent stability under long-term processing conditions, that they can be recovered and continue to exhibit very good performance after multiple reuses. These properties highlight that the ILs have the flexibility to be useful for a variety of downstream chemical processing techniques, and for use in other applications as well.

Copyright © American Institute of Chemical Engineers. All rights reserved.

471708 Investigation of Lignin Streams Generated during Ionic Liquid Pretreatment of Lignocellulosic Biomass

Abstract:

Lignin is a complex three-dimensional amorphous polymer derived from paracoumaryl, coniferyl, and synapyl alcohols, cross-linked together by different a number of linkages. In plant cell walls, lignin fills the spaces between cellulose and hemicellulose, and it acts like a 'glue' that holds the lignocellulose matrix together. Despite years of research, the lignin structure is still not well understood as a function of feedstock and process. The ability of certain ionic liquids (ILs) to dissolve cellulose and/or lignin enables the possibility of commercial lignin upgrading and valorization to improve the overall biorefinery economics. However, to date depolymerization mechanisms of lignin in ILs is not well understood. In this study, the changes in the chemical composition of lignin after pretreatment with IL were investigated. Three different type of biomass, namely switch grass, eucalyptus and pine were utilized in this study. The relative changes in the common β-O-4, β-β and β-5 inter-unit linkages of lignin during pretreatment process were investigated using solution state two dimensional (2D) nuclear magnetic resonance (NMR). The changes in the size and molecular weight distribution of lignin were analyzed using size exclusion chromatography (SEC) and mass spectrometry.

Copyright © American Institute of Chemical Engineers. All rights reserved.

Swelling and Dissolution of Cellulosic Fibers

Mohammad Ghasemi, Marina Tsianou, Paschalis Alexandridis*

Department of Chemical and Biological Engineering
University at Buffalo, The State University of New York (SUNY)
Buffalo, New York 14260-4200, USA

* E-mail: palexand@buffalo.edu

Abstract

The dissolution of cellulose is a critical step for the efficient utilization of this renewable resource as a starting material for the synthesis of high value-added functional polymers. Important aspects of cellulose dissolution such as solvent-induced decrystallization and chain disentanglement are not well-understood. We have developed a phenomenological model that captures the phenomena governing the dissolution of semicrystalline polymers. This model allows the quantification of two important features, i.e., solvent effectiveness in cellulose (i) decrystallization and (ii) chain disentanglement, the balance of which controls the mechanism and kinetics of cellulose dissolution. The insights obtained from this study improve the understanding of cellulose-solvent interactions underlying decrystallization and disentanglement and their contributions in controlling the kinetics of cellulose swelling and dissolution.

Introduction

The processing of cellulose into desired products typically involves its swelling and/or dissolution.[1] A major barrier to cellulose solubility is the recalcitrance of cellulose, which originates from a partial crystalline structure and extended noncovalent interactions.[2] Only a few solvent systems are effective for direct dissolution of cellulose, and they operate under rather strict conditions of composition and temperature.[3] It is generally thought that the capacity of the solvent to break the intra- and inter-molecular hydrogen bond network of cellulose is a key factor in its dissolution.[4] However, it was recently suggested that, since cellulose has amphiphilic properties, hydrophobic interactions might play a significant role in cellulose solubility.[5]

The homogenous swelling and dissolution of cellulose fibers is shown schematically in Figure 1. When solvent diffuses into the cellulose crystalline domains they are gradually disrupted become amorphous, and the fiber swells.[6] After a certain induction time, cellulose chains gradually become disentangled and move out of their network into the bulk solution.[7] Therefore, the dissolution of cellulose involves two inter-related phenomena: decrystallization and chain disentanglement.[7] If a solvent can destroy the crystalline network of cellulose but is not capable to disentangle the chains, then the crystallinity of cellulose is reduced, however, no dissolution happens, as in the case of 1-N-butyl-3-methylimidazolium chloride ([bmim]Cl) at 70 °C.[8] On the other hand, if a solvent can only disentangle the chains, then only the amorphous regions of cellulose can dissolve.

In order to gain insights into the swelling and dissolution of cellulose, a microscopic description of macroscopic experimental information has been provided by using a newly developed phenomenological model [7] that captures the phenomena governing the dissolution of semicrystalline cellulose. Our model fits well experimental data [8] for swelling and dissolution of cotton fibers in the ionic liquid [bmim]Cl.

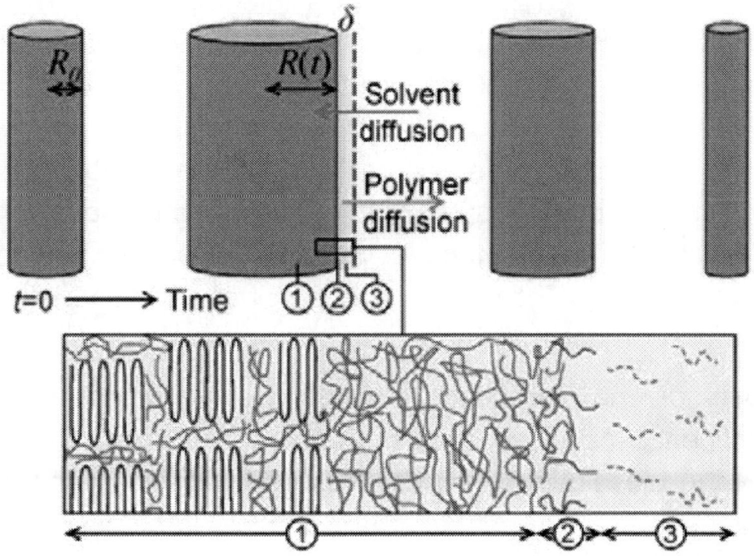

Fig. 1: *Schematic representation of swelling and dissolution of cellulose fiber: (1) fiber interior: crystalline domains (black) and amorphous domains (blue, light shade); (2) fiber-solvent interface at which disentanglement occurs; (3) external diffusion boundary layer at which disentangled cellulose chains (red and dashed) diffuse toward the bulk solution.*[7]

Methods

In the model the cellulose fiber is considered as a long semicrystalline polymer cylinder (consisting of two components; crystalline and amorphous) with a uniform cross section area and with properties that may vary along the cross section.[7] The model describes different steps involved in cellulose dissolution, e.g., solvent diffusion, transformation from crystalline to amorphous domains, specimen swelling, cellulose chain disentanglement, and diffusion of disentangled chains toward the bulk solution, using appropriate transport phenomena and kinetics equations.[7] The ability of solvent molecules to transform the crystal lattice into amorphous and to disentangle cellulose chains is captured by the decrystallization rate constant (k_{dec}) and the disentanglement rate (r_{dis}), respectively. These two are the only parameters fitted to experimental data; all other model parameters are independently obtained.

Results and Discussion

The impact of cellulose decrystallization in the dissolution process is not the same at all decrystallization constant values: at some conditions decrystallization has a strong effect, while at other conditions its effect is negligible (Figure 2a). For example, decreasing k_{dec} from 2.5×10^{-1} to 2.5×10^{-3} s^{-1} has little impact on the total dissolution time (obtained when fraction of dissolved fiber = 1), however, decreasing k_{dec} below 2.5×10^{-3} s^{-1} has a significant impact on the time evolution of fiber dissolution. For $k_{dec} \leq 2.5\times10^{-4}$ s^{-1}, the contribution of solvent molecules to the decrystallization of cellulose is less than their contribution in dissociating the chains, and so the dissolution process is mainly controlled by decrystallization. Because of this, a small change in the k_{dec} value results in a considerable variation in the time evolution of the fiber radius, crystallinity, and solubility, whereas for $k_{dec} \geq 2.5\times10^{-3}$ s^{-1} these fiber characteristics are not affected much by k_{dec}.

When the cellulose-solvent interactions that affect chain disentanglement are stronger than those that influence cellulose decrystallization, the fiber remains crystalline up to almost the end of the dissolution process. This condition happens for $r_{dis} \geq 1\times10^{-9}$ m/s (Figure 2b). Therefore, for $r_{dis} \geq 1\times10^{-9}$ m/s, solvent molecules can disentangle the amorphous domains well, while the decrystallization rate is not high enough to transform the crystalline domains at the fiber surface to amorphous (decrystallization control). However, when $r_{dis} \leq 5\times10^{-10}$ m/s the decrystallization rate of cellulose is relatively greater than the rate of disentanglement, and so the overall kinetics of fiber dissolution are mainly controlled by the chain disentanglement.

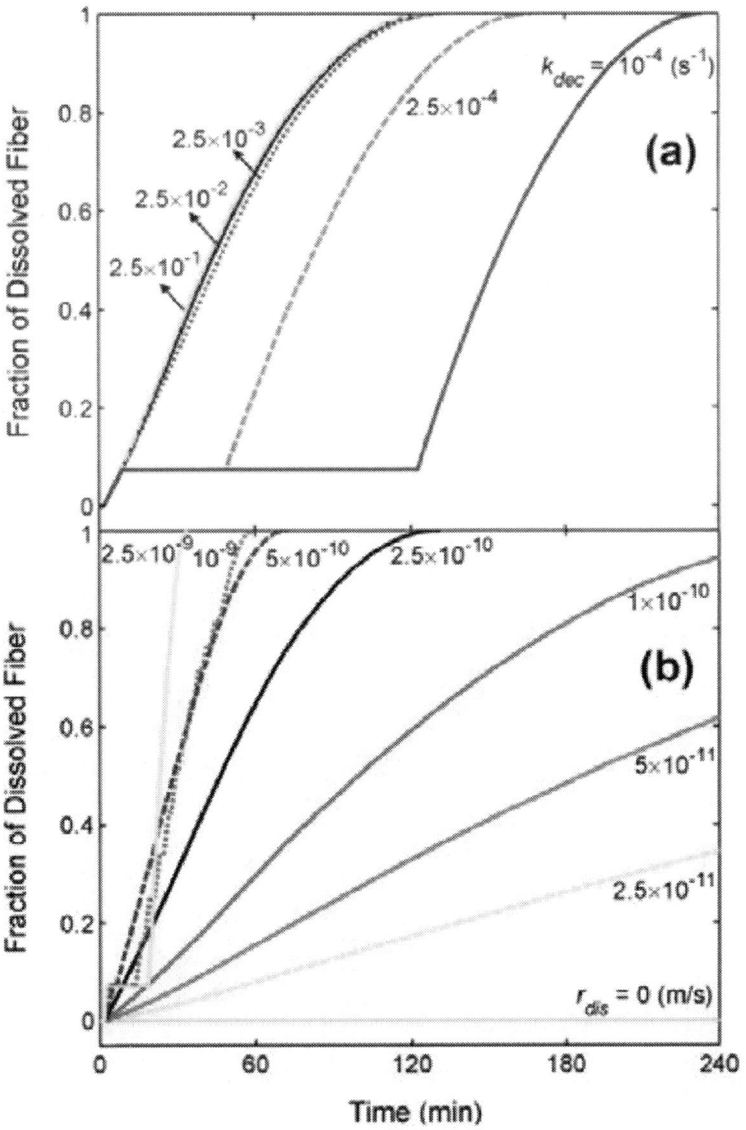

Fig. 2: *Effect of (a) k_{dec} and (b) r_{dis} on the time evolution of the fraction of the fiber that has dissolved.*[7]

The above approach allows us to evaluate the conditions of cellulose dissolution in a variety of systems in the context of the relative contributions of decrystallization and disentanglement. Accordingly, the dissolution conditions can be organized into four regimes: (1) $k_{dec} \leq 2.5\times10^{-4}$ 1/s and $r_{dis} \leq 5\times10^{-10}$ m/s. At these conditions the ability of solvent in both cellulose decrystallization and chain disentanglement is not good, the solubility is limited and the process can be controlled by either decrystallization or disentanglement, depending on the k_{dec} and r_{dis} values. (2) $k_{dec} \leq 2.5\times10^{-4}$ 1/s and $r_{dis} \geq 1\times10^{-9}$ m/s. Here the cellulose-solvent interactions which affect disentanglement are much stronger than interactions that impact decrystallization (decrystallization control).

(3) $k_{dec} \geq 2.5\times10^{-3}$ 1/s and $r_{dis} \leq 5\times10^{-10}$ m/s. At this condition the solvent is more effective in disturbing cellulose intermolecular interactions that result in decrystallization than the interactions that disentangle the chains (disentanglement control). (4) $k_{dec} \geq 2.5\times10^{-3}$ 1/s and $r_{dis} \geq 1\times10^{-9}$ m/s. Here the rates of both decrystallization and disentanglement are very high, the solvent is very aggressive toward cellulose dissolution, and the mechanism of the dissolution process can be controlled by either decrystallization or disentanglement, depending on the k_{dec} and r_{dis} values.

Conclusions

When the solvent effectiveness in the disruption of cellulose crystals is relatively lower than its ability to disentangle the chains (e.g., for LiCl/DMAc[9]), the kinetics of dissolution are controlled by decrystallization. Furthermore, conditions that facilitate cellulose chain disentanglement, in addition to increasing the rate of dissolution, can result in faster decrystallization. The solvent effectiveness in chain disentanglement is the only factor that determines the decrease of the cellulose fiber radius. In cases where the fiber dissolution rate is higher than the decrystallization rate (e.g., in [bmim]Cl at 80 °C[8]), the dissolution of cellulose is mostly controlled by the solvent ability to disentangle the chains. Such knowledge allows the construction of operating condition diagrams for the design of dissolution processes in the biorefinery.

Acknowledgements

We thank the U.S. National Science Foundation (Grant CBET-1159981) for supporting research on cellulose dissolution in our laboratory.

References

(1) Wang, H.; Gurau, G.; Rogers, R. D., Ionic liquid processing of cellulose. *Chem. Soc. Rev.* **2012,** *41* (4), 1519-1537.

(2) Himmel, M. E.; Ding, S.-Y.; Johnson, D. K.; Adney, W. S.; Nimlos, M. R.; Brady, J. W.; Foust, T. D., Biomass recalcitrance: Engineering plants and enzymes for biofuels production. *Science* **2007,** *315* (5813), 804-807.

(3) Liebert, T., Cellulose solvents - remarkable history, bright future. *ACS Symp. Ser.* **2010,** *1033*, 3-54.

(4) Keshk, S. M. S., Effect of different alkaline solutions on crystalline structure of cellulose at different temperatures. *Carbohydr. Polym.* **2015,** *115*, 658-662.

(5) Medronho, B.; Romano, A.; Miguel, M. G.; Stigsson, L.; Lindman, B., Rationalizing cellulose (in)solubility: Reviewing basic physicochemical aspects and role of hydrophobic interactions. *Cellulose* **2012,** *19* (3), 581-587.

(6) El Seoud, O. A., Understanding solvation. *Pure Appl. Chem.* **2009,** *81* (4), 697-707.

(7) Ghasemi, M.; Alexandridis, P.; Tsianou, M., Cellulose dissolution: Insights on the contributions of solvent-induced decrystallization and chain disentanglement. *Cellulose* **2016**.

(8) Jiang, G.; Huang, W.; Wang, B.; Zhang, Y.; Wang, H., The changes of crystalline structure of cellulose during dissolution in 1-butyl-3-methylimidazolium chloride. *Cellulose* **2012,** *19* (3), 679-685.

(9) Ramos, L. A.; Morgado, D. L.; Gessner, F.; Frollini, E.; El Seoud, O. A., A physical organic chemistry approach to dissolution of cellulose: Effects of cellulose mercerization on its properties and on the kinetics of its decrystallization. *Arkivoc* **2011,** *7*, 416-425.

471942 Towards a New Paradigm in the Ionic Liquid Pretreatment for the Production of Lignocellulosic Biofuels: Technoeconomic Insights

Abstract:

Ionic liquids (ILs) have been receiving increasing attention in biomass pretreatment applications (Konda et al., 2014 and 2015) as some ionic liquids are shown to be very efficient in biomass pretreatment. Among several benefits of ionic liquids in general, they are often regarded as *green solvents* due to their negligible volatility. More importantly, given the availability of numerous anions and cations, it is often possible to *tune* their properties to suit specific application(s) by appropriately selecting the combination of anion and cation. Specific to biomass pretreatment, some ionic liquids are shown to be feedstock agnostic (Li et al., 2013) which is a very desirable characteristic especially given the uncertainty with feedstock supply and quality (e.g., seasonal variations). Such a technology can be potentially disruptive and can play a critical role in enabling large-scale cellulosic biorefineries.

To tap into the full potential of ionic liquids, however, it is important that the processes are designed such that they are economically viable as well as environmentally sustainable. These criteria demand for minimal/reduced use of any other chemicals (e.g., co-solvents, water) and energy footprint in the entire process. Furthermore, given the highly integrated nature of these cellulosic biorefineries, these pretreatment technologies need to be developed in such a way that they can be â€˜*plugged-in*â€™ while minimizing any negative impact on the other sections of the process (including downstream). Realizing such a process is often challenging as there exists significant trade-offs amongst these criteria. This emphasizes the need for greater insights into the mass and energy balances of such complex processes. To this end, with the help of virtual biorefinery models, technoeconomic analysis (TEA) offers invaluable guidance to nurture technological advances. Equipped with technoeconomic insights, scientists at Joint BioEnergy Institute (JBEI) are harnessing the tunable nature of ILs, together with the novel process design and operational concepts, to develop advanced pretreatment technologies. Our previous TEA work (Konda at al., 2014) emphasized the importance of high biomass loading, high IL recovery, low IL and water usage to ensure the prospective cellulosic biorefineries can indeed be economically sustainable. A recent advancement in this direction includes the development of a novel integrated process that is capable of high yields under high-gravity (HG) conditions, while minimizing the use of water and IL. TEA revealed that the continued developments with the proposed integrated process could potentially reduce minimum fuel selling price (MFSP) to below $3/gal.

References:

- Konda et al. (2014). Understanding cost drivers and economic potential of two variants of ionic liquid pretreatment for cellulosic biofuel production. Biotechnology for Biofuels 7: 86.
- Konda et al. (2015). Towards a cost-competitive ionic liquid based pretreatment: Insights from technoeconomic analysis. AIChE Annual Meeting, November 2015, Salt Lake City, UT.
- Li et al. (2013). Comparing the recalcitrance of eucalyptus, pine, and switchgrass using ionic liquid and dilute acid pretreatments." BioEnergy Research 6, no. 1 (2013): 14-23.

Copyright © American Institute of Chemical Engineers. All rights reserved.

469060 Estimating the Life Cycle Impact of Chemicals from Molecular Descriptors and Thermodynamic Properties Via Mixed-Integer Linear Programming

Abstract:

The chemical industry is at present striving to decrease its environmental footprint during the transition towards a more sustainable manufacturing sector. In this context, there is a clear need to develop systematic methods to assess and optimize the environmental impact of chemicals, identify critical hotspots across their supply chains and define the best strategies to implement so as to enhance their sustainability level.

In the recent past, it has become clear that the environmental impact should be assessed considering all the stages in the life cycle of the products/services being assessed. As a result, Life Cycle Assessment (LCA) has become (arguably) the prevalent approach to quantify the environmental burdens of products, finding applications in a wide variety of areas, including the evaluation of the environmental footprint of chemicals from cradle to grave.

One of the main limitations of LCA is that it is highly data intensive, that is, it requires large amounts of data from several echelons of the product supply chain, which are typically owned by different companies that might be reluctant to share this environmental information or even lack the corresponding measurements. This is critical in the chemical industry, where chemicals are produced by complex interrelated chemical networks that exchange energy, mass and water and consume a wide range of intermediate products and feedstocks.

One possible way to simplify the LCA calculations is to resort to Streamlined LCA (SLCA) approaches, which attempt to reduce the upstream and downstream information required in a standard LCA by using proxy data, qualitative models and/or regression equations. These methods are typically tailored to a particular sector, as otherwise the underlying simplifications and assumptions would not hold, thereby leading to poor estimates and large errors.

SLCA methods have been developed for a wide range of production processes, including the building sector (where the size of the bill of materials often makes unpractical the use of standard approaches), urban water treatments and hydroelectric power plants, among others. In the context of chemical processes, SLCA has been applied in oil refining as well as to general chemical processes.

In this work we introduce a novel approach for evaluating the life cycle impact of chemicals that provides better estimates and reduces the complexity of previous models, thereby opening new avenues for their use within the framework of computer aided molecular design tools. In essence, building on the seminal paper by Wernet et Al. (2008)[1] that predicted the life cycle impact of chemicals from their molecular structure using neural networks, we present here an alternative method based on rigorous mixed-integer optimization techniques that uses key thermodynamic properties in addition to molecular descriptors in order to produce more accurate estimates of impact. Our methodology was applied to estimate 9 environmental impact categories of 88 chemicals using 17 structural descriptors and 15 thermodynamic properties. Numerical results show that the life cycle impacts can be predicted with relative errors ranging from 18 to 45% depending on the specific indicator. Furthermore, the inclusion of thermodynamic properties in addition to molecular descriptors improves greatly the quality of the estimate, with relative errors dropping from 22 to 18% when moving from a molecular-based to a combined-based estimation.

1 G. Wernet, S. Hellweg, U. Fischer, S. Papadokonstantakis and K. Hungerbuĭ˜hler, *Environ. Sci. Technol.*, 2008, **42**, 6717â€"6722.

Copyright © American Institute of Chemical Engineers. All rights reserved.

462465 Optimal Production of Light Olefins from Lignocellulosic Biomass (BTO): Process Synthesis and Global Optimization

Abstract:

The growing interest toward alternative carbonaceous energy sources stems from concerns surrounding energy independency, energy affordability, and the generation of lower carbon energy, as stated in a recent perspective [1]. Specifically, there is a need for the development of sustainable and carbon neutral energy processes that have the potential to reduce global energy related carbon dioxide emissions. Lignocellulosic biomass is a potential feedstock that can serve as an alternative to fossil-fuels and offset the growing energy demand while simultaneously reducing the net carbon emissions due to the absorption of CO_2 during photosynthesis. The U.S. has the potential of sustainably supplying more than one billion tons of biomass annually [2,3]. Biomass has already been shown to be a viable feedstock for the production of liquid transportation fuels at low break-even oil prices [4]. However, the transportation fuel demand is expected to stall in the near future while the chemicals demand is expected to raise by 65% [5]. This increase in the chemicals demand is expected to be the major driver of the crude oil demand growth [5].

This work investigates the optimal production of light olefins from biomass within a process synthesis and global optimization framework. Light olefins are the most demanded class of petrochemicals and command two-thirds of the petrochemical commodities market [6]. These chemicals can be produced from carbonaceous feedstocks through several novel/competing process alternatives [7]. We propose a process superstructure with alternatives regarding (i) biomass conversion, (ii) synthesis gas cleaning, (iii) methanol production and conversion, (iv) olefins purification, (v) light gas recycle, and (vi) hydrogen and oxygen production among others. Simultaneous heat, power, and water integration is included to ensure minimum utility usage. The objective function of the proposed mathematical model maximizes the profit of the refinery. The resulting large-scale mixed integer nonlinear optimization model (MINLP) is solved to global optimality with a branch-and-bound global optimization framework. Several case studies are investigated to observe the effect of biomass type, plant scale, and topological decisions on the overall plant profitability. Significant reductions in the lifecycle GHG emissions are observed through the utilization of biomass.

[1]: Floudas, C. A.; Niziolek, A. M.; Onel, O.; Matthews, L. R. Multi-scale systems engineering for energy and the environment: Challenges and opportunities. AIChE Journal 2016, 62 (3), 602-623.

[2]: Department of Energy, Biomass as Feedstock for a Bioenergy and Bioproducts Industry: The Technical Feasibility of a Billion-Ton Annual Supply.

[3]: Department of Energy, US Billion-Ton Update: Biomass Supply for a Bioenergy and Bioproducts Industry.

[4]: Baliban, R. C.; Elia, J. A.; Floudas, C. A. Biomass to liquid transportation fuels (BTL) systems: process synthesis and global optimization framework. Energy Environ. Sci. 2013, 6 (1), 267-287.

[5]: ExxonMobil, The Outlook for Energy, A View to 2040

[6]: de Klerk, A. Fischer-Tropsch Refining; Wiley-VCH Verlag & Co. KGaA, 2011.

[7]: Onel, O.; Niziolek, A. M; Floudas, C. A Optimal Production of Light Olefins from Natural Gas via the Methanol Intermediate. Industrial & Engineering Chemistry Research 2016, 55 (11), 3043-3063.

Copyright © American Institute of Chemical Engineers. All rights reserved.

455402 Sub and Supercritical Conversion of Kirchneriella Algal Biomass into Biocrude Oil

Abstract:

Sub and supercritical water technology utilizes wet biomass as feedstock to produce liquid fuels. Water at elevated temperature and pressure exhibit extraordinary solvent properties. Here, water acts as reactant as well as reaction medium in performing various complex chemical reactions. Algal biomass was an interesting feedstock for producing biofuel via sub and supercritical technology due to their high water content. We investigated the potential of microalgae (Kirchneriella sp.) for biofuel production. Several process parameters such as reaction temperature, pressure, time and solid loading were studied. Biocrude oil yield of 45.5% was achieved at 300°C, 30 min, 100 bar and 10% solids loading. Lower biocrude oil was produced at supercritical conditions (374°C and 220 bar) compared to subcritical conditions due to intensive gasification at higher temperatures. The high heating values (HHV) of biocrude oil and bio-char were determined with a bomb calorimeter. The biocrude oils and water soluble organics were Analyzed using gas chromatography mass spectrometry (GC-MS) and calorimetrically. CHNS/O analyzer was used to determine the elemental composition of algae, biocrude and biochar samples.

Copyright © American Institute of Chemical Engineers. All rights reserved.

471483 Selective Aromatic Ring-Opening of Biorefinery Lignin Towards Dicarboxylic Acids:Â Influence of Reaction Temperature and pH

Abstract:

Lignin is the worldâ€™s only sustainable source of aromatic compounds, which has resulted in a significant quantity of research being conducted on the depolymerization of this complex biopolymer for the production of low molecular weight aromatic compounds. These aromatic products can then be further broken down via selective ring-opening to form dicarboxylic acids.

Dicarboxylic acids are a valuable class of organic compounds which are utilized in numerous commercial industries, including the production of biopolymers, pharmaceuticals, and food products. Like phenolic compounds, dicarboxylic acids are currently dependent on crude oil as a feedstock. However, lignin has the potential to act as a sustainable alternative.

This work investigates the selective production of dicarboxylic acids from several different types of lignins, via thermochemical breakdown. The treatment conditions employed a temperature range of 60 to 250 Â°C, and a reaction mixture pH of 4 to 13. The major products obtained in this work were oxalic, malonic and succinic acids. Yields of up to 10 wt.% oxalic acid were achieved, based on the mass of lignin treated.

Copyright © American Institute of Chemical Engineers. All rights reserved.

467791 Lignin in Ethylene Glycol and Poly(Ethylene Glycol): Fortified Lubricants with Internal Hydrogen Bonding

Abstract:

Lignin, one of the most naturally abundant polymers, has been successfully incorporated into ethylene glycol (EG) and polyethylene glycol (PEG) in this work and fortified lubricating properties were achieved in EG/lignin and PEG/lignin. The molecular interaction between lignin and EG (or PEG) has been revealed as hydrogen bonding, which serves as the dominating factor that determines the thermal, rheological and tribological properties of the mixed systems of EG/lignin and PEG/lignin. The physicochemical properties of the mixed lubricants are tightly related to the state of internal hydrogen bonding (EG-EG, PEG-PEG, EG-lignin, PEG-lignin and lignin-lignin) and are well correlated to their lubrication properties. Generally, larger lignin fraction leads to better lubricating performance in both EG and PEG systems. Lignin liquefaction in PEG has been addressed by catalytic degradation with the presence of sulfuric acid, which was then neutralized by triethanolamine for lubricant development. Lignin in PEG significantly improves the lubricating property at higher pressure conditions, where a wear reduction of 94.6% was observed. Lignin fortified EG and PEG based lubricants show outstanding non-corrosive characteristic to the mostly used metal materials such as aluminum and iron.

Copyright © American Institute of Chemical Engineers. All rights reserved.

469693 Enzyme Selection for Hydrolysis of Lignocellulosic Biomass Coupled with Fermentation

Abstract:

The India - Norway collaborative project EcoLodge aims at developing a new, integrated biotechnological production process for butyl butyrate, a promising supplement/substitute for diesel and jet engine fuels. The feedstock, lignocellulosic biomass is enzymatically hydrolyzed to produce C5 and C6 sugars. Further, the sugars are fermented by the bacteria to produce butanol and butyric acid, which are the building blocks of butyl butyrate. Butanol and butyric acid fermentations are developed and optimized separately to achieve high yield.

Major challenges for fermentation of biomass derived sugars are low product yield and high hydrolysis cost (Sun and Cheng, 2002). Exploitation of all the C5 and C6 sugar polymers locked in the lignocellulosic biomass can provide a solution to the low yield problem. However, the hydrolysis of all the polymers requires an optimal enzyme mixture, which is usually need to be tailor made or carefully selected for the specific substrate (Jørgensen et. al., 2007). The challenge of high enzymatic hydrolysis cost is mainly due to the enzyme since utility costs are low owing to mild operating conditions (pH 4.8 and temperature 45 - 50 °C) (Duff and Murray, 1996). Therefore, this paper suggests an optimal enzyme mixture, which yields the most favorable composition of C5 and C6 sugars necessary for high yield butanol and butyric acid fermentations.

Mathematical models and kinetic parameters for bacterial growth on mixed substrates from the literature are employed for simulation of fermentations with varied sugar compositions (Lendemann and Egli, 1998). Various composition values of C5 and C6 sugar mixtures are taken from the literature and used in the model simulations as well for comparison purposes (Van Dyk and Pletschke, 2012). The simulation results illustrate the effect of sugar composition on the fermentation yield; eventually provide the optimal value of the sugar composition and corresponding enzyme mixtures. The approach enables backward knowledge flow, from fermentation to enzymatic hydrolysis, to investigate the optimal process conditions. Therefore, it suggests the optimal enzyme mixture after evaluating the possibilities in terms of the cost, composition and the operating conditions.

References

Sun, Y., & Cheng, J. (2002). Hydrolysis of lignocellulosic materials for ethanol production: a review. *Bioresource technology*, *83*(1), 1-11.

Jørgensen, H., Kristensen, J. B., & Felby, C. (2007). Enzymatic conversion of lignocellulose into fermentable sugars: challenges and opportunities.*Biofuels, Bioproducts and Biorefining*, *1*(2), 119-134.

Duff, S. J., & Murray, W. D. (1996). Bioconversion of forest products industry waste cellulosics to fuel ethanol: a review. *Bioresource technology*,*55*(1), 1-33.

Lendenmann, U., & Egli, T. (1998). Kinetic models for the growth of Escherichia coli with mixtures of sugars under carbon-limited conditions.*Biotechnology and bioengineering*, *59*(1), 99-107.

Van Dyk, J. S., & Pletschke, B. I. (2012). A review of lignocellulose bioconversion using enzymatic hydrolysis and synergistic cooperation between enzymesâ€"factors affecting enzymes, conversion and synergy.*Biotechnology advances*, *30*(6), 1458-1480.

Copyright © American Institute of Chemical Engineers. All rights reserved.

461802 Performance of a Bench-Scale Continuous Hydrothermal Carbonization Reactor

Abstract:

Hydrothermal carbonization (HTC, thermal hydrolysis, or wet torrefaction) is a treatment process which converts moist feedstocks into homogenized, carbon rich, and energy dense solid fuel, called hydrochar. One of the main advantages of HTC compared to other thermochemical treatment processes is the use of moisture as reaction medium and reactant, thereby precluding the need for drying prior to treatment. Thermodynamic properties of water change greatly in the subcritical region from 180-280 °C, and as a result, subcritical water behaves as a non-polar solvent and mild acid and base catalyst simultaneously. When subjected to HTC biomass undergoes rapid hydrolysis and other reactions, releasing oxygen-containing volatiles and producing a highly hydrophobic hydrochar.

Although HTC offers a relatively simple and straightforward solution to process diverse biomass feedstocks, the requirements of high pressure and high temperature make the process complex and potentially costly to design and operate. The lab-scale batch process has already been demonstrated in various laboratories around the world, but batch process is not cost-effective for industrial-scale deployment. A continuous process offers a relatively small footprint, energy recovery and efficiency, and notably, economies of scale.

In this study, a bench-scale continuous HTC reactor system has been designed, constructed, commissioned, and operated with model compounds and with whole biomass feedstocks. The maximum temperature and pressure of the reactor are 250 °C and 40 bar. The throughput of the reactor system was maintained at 0.3 L/min with solids loading of ~5%, while the reaction time was maintained at 5 min. The basic reactor configuration is a section of straight pipe, and numerous engineering challenges were encountered during design and commissioning. The primary challenges are related to instrumentation durability, temperature control, and flow control.

Results are presented for conversion of glucose and microcrystalline cellulose. Both solid and liquid products were tested for their physico-chemical properties and compared with the corresponding products from batch process generated in a 2L batch Parr reactor. It was found that temperature and pressure were stable during operation and products are relatively similar to that of batch process.

Copyright © American Institute of Chemical Engineers. All rights reserved.

474623 Spectroscopic and Thermal Characterization of Carbonized Food Wastes

Abstract:

Waste-to-energy conversion technologies may play an important role to improve the efficiency and sustainability of global food supply. In this worked, performed in collaboration with researchers at UNICAMP (Brazil), we used thermal and spectroscopic methods to characterize the carbonized residues that remain after flow-through hydrothermal extraction and hydrolysis. A recent energy analysis of flow-through hydrothermal hydrolysis finds that it may be energetically favorable compared to other pre-treatment and saccharification techniques. Three factory wastes produced by food industries were studied: sugarcane bagasse, green coffee powder, and defatted coffee cake. We foucs here on results obtained for the coffee wastes. Thermogravimetric analysis (TGA) was used to monitor extraction of semi-volatile components after treatment at 150 °C, followed by hydrolysis and carbonization as temperatures were increased to 250 °C. Infrared spectroscopy indicated that the extracted semi-volatile compounds were a combination of valuable lipids and organic acids. At higher temperatures, infrared spectroscopy identified preferential hydrolysis of hemicellulose components, with cellulose and lignin remaining. Raman microscopy gave further detail, identifying the specific compound chlorogenic acid, a valuable compound marketed as a nutraceutical, as one of the semi-volatile compounds to be extracted from the coffee. At higher temperatures, Raman microscopy provided evidence of carbonization of the coffee waste, as suggested by the appearance of graphitic and defect bands previously associated with hydrothermal char. As shown in the Figure, our results suggest temperature-programmed extraction, hydrolysis, and carbonization of coffee wastes to produce biorefinery streams consisting of valuable organic compounds, bioenergy precursors, and carbonized char.

Copyright © American Institute of Chemical Engineers. All rights reserved.

474753 Drying and Pyrolysis of Solid Waste on Spacecraft for Water Recovery and Biochar

Abstract:

Manned spaceflight outside of low-Earth orbit will require significant advances in closing loops within life support systems, especially the recycling of solid and liquid wastes to produce oxygen, food, and fresh water. Pyrolysis is one waste conversion method that has received considerable attention due to its potential to enable improvements in waste reduction, material recovery, and energy production.

Here, we describe the development of a slow pyrolysis reaction system that can transform solid waste and brine from the water treatment system into a nutrient-rich crop growth medium, while recovering water and carbon dioxide. The feedstock for the process followed a standard s recipe used by NASA to represent the average daily solid waste produced by astronauts on manned missions. This recipe included packaging materials (polyethylene, nylon, aluminum foil), clothing and towels (mostly cotton), personal care products (i.e. shampoo, toothpaste, facial tissue, hand wipes, diapers, etc.), food waste, metabolic waste simulant, and miscellaneous materials such as nitrile gloves, duct tape, and cleaning wipes. Solid waste processing consisted of two steps: drying to recover water and pyrolysis at moderate temperatures (400-550°C). Process steps were optimized to recover the most water and produce the most suitable biochar for plant growth applications within the assumed 8 hour total operational time window. Recovered water and biochars were characterized using a variety of physical, chemical and soil nutrient analysis methods. We discuss the challenges of working with this waste feedstock compared to lignocellulosic materials, how this system fits into an overarching life support architecture, and future work needed to adapt this system to a microgravity environment.

Copyright © American Institute of Chemical Engineers. All rights reserved.

474765 Evaluating Physical and Electronic Structure Developments of Loblolly Pine Derived Biochar and Activated Carbon

Abstract:

Biochar is a major byproduct of fast pyrolysis process. Until recently, biochar is considered as an inexpensive solid fuel that is re-consumed for the pyrolysis reactor heating. However, advances in nanotechnologies reveal the potential high-end applications of biochar as electrodes for electrochemical energy storage systems and conductive thin film materials for sensors and electronic devices. The high-end applications of biochar not only contribute to the profitability of overall fast pyrolysis process, but also contribute to the growth of sustainable material market. To attain the objective, it is imperative to understand the complex biomass derived carbon structural development during the carbonization and the activation. In this study, the physical and the electronic structures of loblolly pine (*Pinus Taeda*) derived biochar and activated carbon were systematically analyzed by interpreting X-ray diffraction patterns, electron energy loss spectra, and electrical conductivities. Loblolly pine was carbonized at nine different temperatures (300 , 350 , 400 , 500 , 600 , 700 , 800 , 900 , and 1,000 ,) using quartz tube furnace. Then, four biochar samples carbonized at 300 , 350 , 500 , and 700 were impregnated with NaOH to produce the activated carbon. Carbon sp^2 content was quantitatively determined by the electron energy loss spectroscopy. The sp^2 content of biochar produced at 300 is 58% and the sp^2 content gradually increases to 78% as the carbonization temperature is elevated to 1,000 . The sp^2 contents of activated carbon are 74%, 78%, 79%, and 85% where precursor biochar carbonization temperatures are 300 , 350 , 500 , and 700 . Polyaromatic cluster size and graphitic stacking thickness were determined by analyzing the X-ray diffraction patterns. Four stages of biochar physical structural development were determined. After major biomass components decompose, the average polyaromatic cluster size and the graphitic stacking thickness increase as the carbonization temperature increases. Carbon paste electrode was produced by mixing biochar and activated carbon powder with polyvinylidene fluoride solution. Biochar electrodes features were closer to insulator that the electrical conductivity was not measurable. Activated carbon electrodes had good electrical conductivities which values are 1.14 S/cm, 1.41 S/cm, 1.3 S/cm, and 1.58 S/cm. The detailed model of physical and electronic structural developments will be discussed.

Copyright © American Institute of Chemical Engineers. All rights reserved.

466364 Characterization of Free Radicals By Electron Spin Resonance Spectroscopy in Biochars from Pyrolysis at High Heating Rates and at High Temperatures

Abstract:

Understanding fast pyrolysis of biomass-derived materials is an important step in optimization of combustion processes. Similar to coal combustion, the fuel burn out is known to be influenced by the yield and reactivity of chars, produced during pyrolysis. The rapid heating of small biomass particles and the short residence time at high temperatures minimize the char yield and increase char reactivity. The differences in chemical composition of organic and inorganic matter between wood and herbaceous biomass affect the operational flexibility of power plants, and increase the complexity of mathematical models that can predict yields, composition and rates of product (char, tar, light gases) formation from fast pyrolysis. The modeling of cross-linking and polymerization reactions in biomass pyrolysis includes the formation of free radicals and their disappearance. Knowledge about these radical reactions is important in order to achieve the high fuel conversion at short residence times. However, little is known about the extent of free radical reactions in pulverized biomass at fast pyrolysis conditions.

The concentration and type of free radicals from the decay (termination stage) of pyrolysis at slow and fast heating rates and at high temperatures (above 1000°C) in biomass char have been studied. A room-temperature electron spin resonance spectroscopy study was conducted on original wood, herbaceous biomass, holocelluloses, lignin and their chars, prepared at high temperatures in a wire mesh reactor, an entrained-flow reactor, and a tubular reactor. The radical concentrations in the chars from the decay stage range up between $7x10^{18}$ and $1.5x10^{19}$ spins g^{-1}.

The results indicated that any differences in the biomass major constituents (cellulose, hemicellulose, lignin) had a minor effect on remaining radical concentrations compared to potassium and silica contents. The higher radical concentrations in the wheat straw chars from the decay stage of pyrolysis in the entrained flow reactor compared to the wood chars were related to the decreased mobility of potassium in the char matrix, leading to the less efficient catalytic effects of potassium on the bond-breaking and radical re-attachments. The high Si levels in the rice husk caused an increase in the char radical concentration compared to the wheat straw because the free radicals were trapped in a char consisting of a molten amorphous silica at heating rates of 10^3°C s^{-1}. The experimental electron spin resonance spectroscopy spectra were analyzed by fitting to simulated data in order to identify radical types, based on g-values and line widths. The results show that at high temperatures, mostly aliphatic radicals ($g = 2.0026$-2.0028) and PAH radicals ($g = 2.0027$-2.0031) were formed. Free radicals structure from the termination stage was modelled by density functional theory (DTF).

Copyright © American Institute of Chemical Engineers. All rights reserved.

474844 Pilot-Scale Catalytic Biomass Pyrolysis Studies

Abstract:

Catalytic fast pyrolysis is the direct liquefaction process used by RTI to produce a hydrocarbon-rich biocrude that can subsequently be upgraded by hydrotreating to make refinable intermediates and â€œdrop-inâ€ advanced biofuels. As part of the technology development, the process has been scaled up from bench-scale to 1 ton per day (1TPD) pilot-scale. Evaluation of the process at the pilot-scale is one of the critical steps being taken by RTI to address a number of technical challenges facing catalytic pyrolysis and to collect required processing and engineering data to support the demonstration and commercialization of the technology.

Up to date, several experimental campaigns have been conducted to optimize the catalytic biomass pyrolysis process to achieve high degree of deoxygenation while maximizing the bio-crude production and improving the biocrude thermal stability. A commercially available spray dried, non-zeolite, alumina-based catalyst has been exclusively studied so far. The quality and biocrude yield has been evaluated at different temperatures and other process conditions. Typically, partially deoxygenated biocrude with oxygen contents between 20 and 25 wt% is produced. Gas Chromatography Mass Spectrometry (GC-MS) analysis have shown high amount of oxygenated aromatics in the partially deoxygenated biocrude. Current results from RTIâ€™s 1 ton per day pilot-scale catalytic pyrolysis unit show a biocrude yield up to 19.2 wt% is achievable. This presentation will be used to discuss in detail the results that have been obtained from several studies and highlight the progress that has been made.

Copyright © American Institute of Chemical Engineers. All rights reserved.

Catalytic Pyrolysis of Bio-Oil Model Compounds over La/Ce/Ni Modified HZSM-5 Zeolites

Fengwen Yu, Qiang SONG, Yong NIE, Zhichao XU, Jianbing JI
Zhejiang University of technology, Hangzhou, China

Abstract :

- With the contradiction between supply and demand of petroleum resources and environmental pollution, more and more attention has been paid to the study of clean and renewable energy, especially converting bio-oil into biofuel. Bio-oil catalytic pyrolysis has several advantages, such as low capital cost, easy handling and safe controlling. It has been widely used in improving the quality of bio-oil, for fossil fuels and chemicals production.
- The experimental present a catalytic strategy to upgrade a simulated bio-oil consisting of ethyl acetate, diacetone alcohol, furfural and guaiacol, aiming to study the effect of La/Ce/Ni modified HZSM-5 zeolites on catalytic pyrolysis. The catalysts properties are investigated by BET and NH_3-TPD and the major constituents of the pyrolytic products are carried out by GC/MS. The results have pointed out the yield of organic compounds is increased from 42.26% to 65.05% by Ce-HZSM-5, whilst keeping lower coke formation rates after adding metal cations under the condition of T=450℃, WHSV=4 h^{-1}. The major components of catalytic pyrolysis are aromatic hydrocarbons, ketones, phenols and so on. The optimal selectivity of aromatic reached 46.37% over Ce-HZSM-5. The La-HZSM-5 improved the selectivity of partial deoxidation products from 4.88% to 21.97% .

2016 AIChE Annual Meeting AIChE

Catalytic pyrolysis of bio-oil model compounds over La/Ce/Ni modified HZSM-5 zeolites

Zhejiang Province Key Laboratory of Biofuel,
College of Chemical Engineering,
Zhejiang University of Technology,
Hangzhou, China

Fengwen YU
Qing SONG
Nie YONG
Zhichao XU
Jianbing JI

CONTENTS

1. Background
2. Experimental
3. Results and discussion
4. Conclusion
5. Acknowledgments
6. References

Part 1

Background

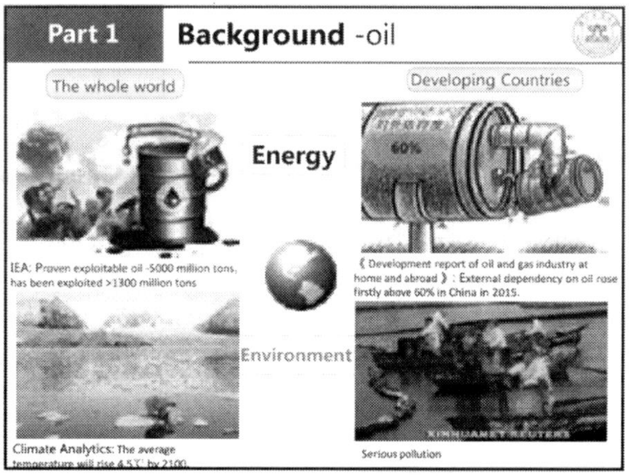
Part 1
Background -oil
The whole world
Developing Countries
Energy
IEA: Proven exploitable oil -5000 million tons, has been exploited >1300 million tons
《Development report of oil and gas industry at home and abroad》: External dependency on oil rose firstly above 60% in China in 2015.
Environment
Climate Analytics: The average temperature will rise 4.5℃ by 2100.
Serious pollution

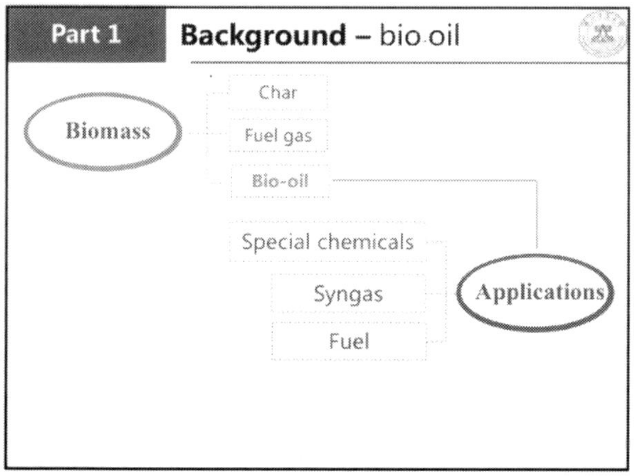
Part 1
Background – bio oil
Biomass
Char
Fuel gas
Bio-oil
Special chemicals
Syngas
Fuel
Applications

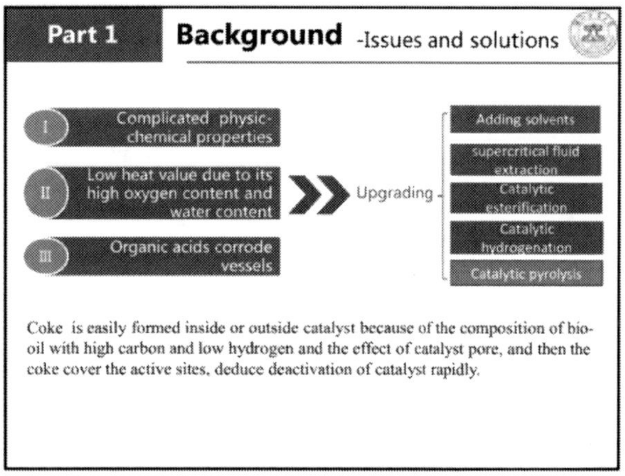
Part 1
Background -Issues and solutions
I Complicated physic-chemical properties
II Low heat value due to its high oxygen content and water content
III Organic acids corrode vessels
Upgrading
Adding solvents
supercritical fluid extraction
Catalytic esterification
Catalytic hydrogenation
Catalytic pyrolysis
Coke is easily formed inside or outside catalyst because of the composition of bio-oil with high carbon and low hydrogen and the effect of catalyst pore, and then the coke cover the active sites, deduce deactivation of catalyst rapidly.

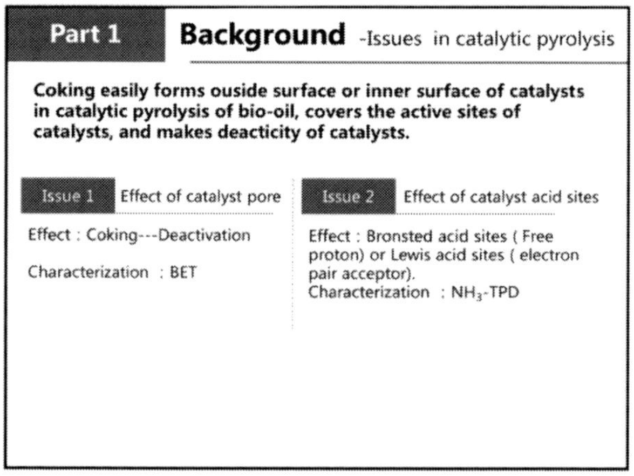
Part 1
Background -Issues in catalytic pyrolysis
Coking easily forms ouside surface or inner surface of catalysts in catalytic pyrolysis of bio-oil, covers the active sites of catalysts, and makes deacticity of catalysts.
Issue 1 Effect of catalyst pore
Effect : Coking---Deactivation
Characterization : BET
Issue 2 Effect of catalyst acid sites
Effect : Bronsted acid sites (Free proton) or Lewis acid sites (electron pair acceptor).
Characterization : NH3-TPD

Part 1 Background –purpose of the research

The key point lies in screening and modifying catalysts in order to obtain high degree of pyrolysis, high effect of deoxidation, and low carbon deposition and coking.

Part 2 Experimental

Part 2 Experimental

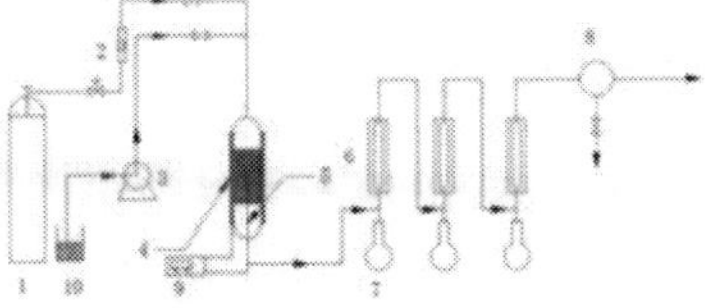

Fig. Experiment flowsheet of catalytic pyrolysis of bio-oil compounds
1-N_2 tank, 2-Rotameter, 3-Peristaltic pump, 4-Heating jacket, 5-Reactor, 6-Condensation pipe, 7-Liquid product receiving bottle, 8-Gas buffer bottle, 9-Automatic temperature control unit, 10-Bio-oil (model compounds with Diacetone alcohol, Ethyl acetate, Furfural, Guaiacol)

- Continuous and convenient to control the temperature

Part 2 Preparation of modified HZSM-5 catalysts

Materials

HZSM-5 (Si/Al=50)
$La(NO_3)_3 \cdot 6H_2O$
$Ni(NO_3)_2 \cdot 6H_2O$
$(NH_4)_2HPO_4$

Catalysts

1.0 wt.%La-HZSM-5,
1.0 wt.%Ce-HZSM-5,
1.0 wt.%Ni-HZSM-5

Procedure

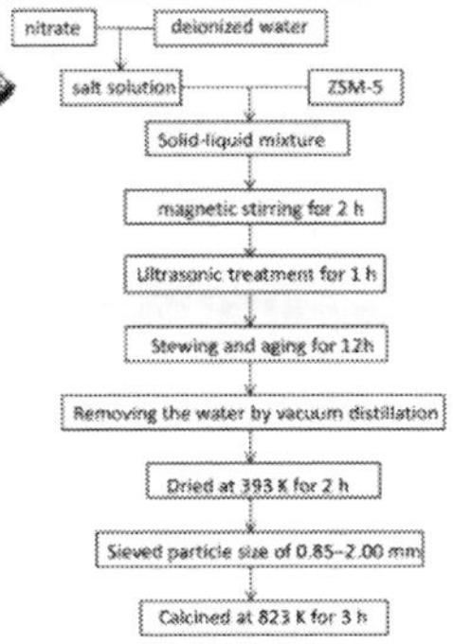

Part 3

Results and discussion

Part 3 Properties of modified HZSM-5 molecular sieve

Table Pore structure parameters of catalysts

Types of Catalysts	$S_{BET}/(m^2 \cdot g^{-1})$	Pore size/nm	Pore volume/ $(ml \cdot g^{-1})$
HZSM-5	308.61	0.77	0.15
La-HZSM-5	300.62	0.79	0.15
Ce-HZSM-5	297.72	0.83	0.15
Ni-HZSM-5	290.42	0.80	0.14

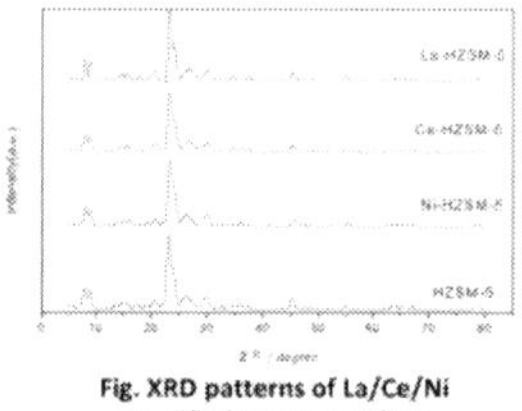

Fig. XRD patterns of La/Ce/Ni modified HZSM-5 zeolites

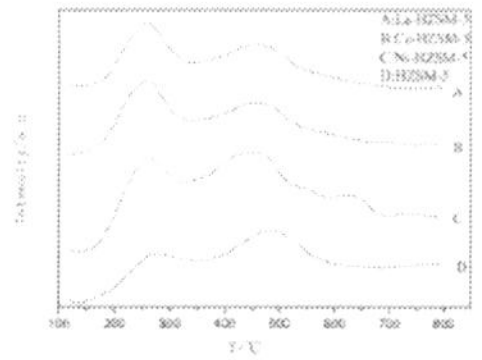

Fig. NH_3-TPD profiles of La/Ce/Ni modified HZSM-5

Part 3 Properties of modified HZSM-5 molecular sieve

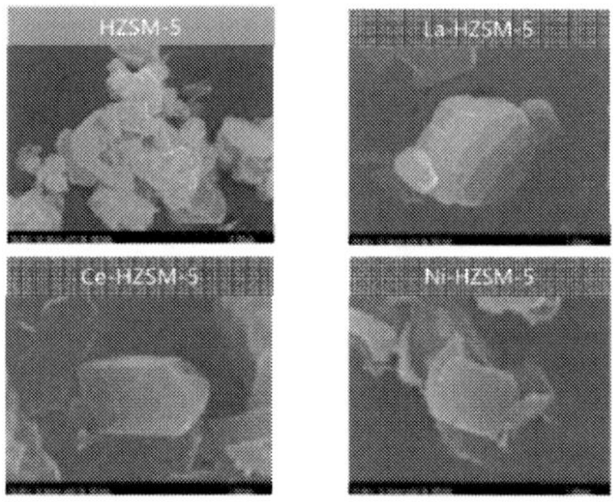

Fig. SEM images of La/Ce/Ni modified HZSM-5 zeolites

Part 3 Effect of temperature on model compounds pyrolysis

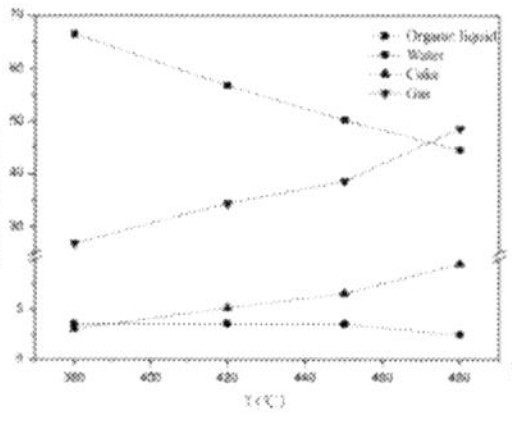

Fig.Products distribution at different temperatures (Feed rate=1.5 ml·min^{-1}, WHSV =4 h^{-1}, no catalyst)

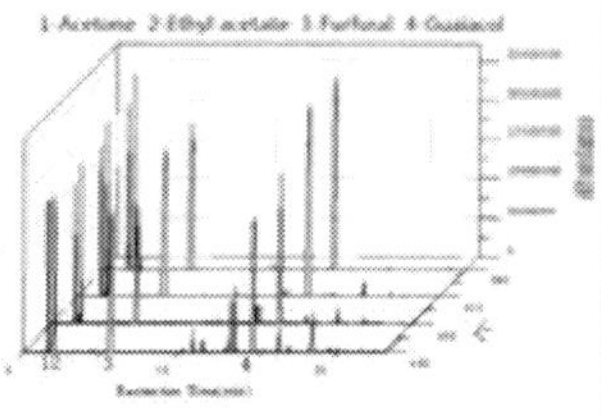

Fig. TIC of organic liquid obtained at different temperatures

- 450℃ is optimal temperature due to higher pyrolysis yield and moderate product distribution.

Part 3 Effect of mdified catalysts on product distribution

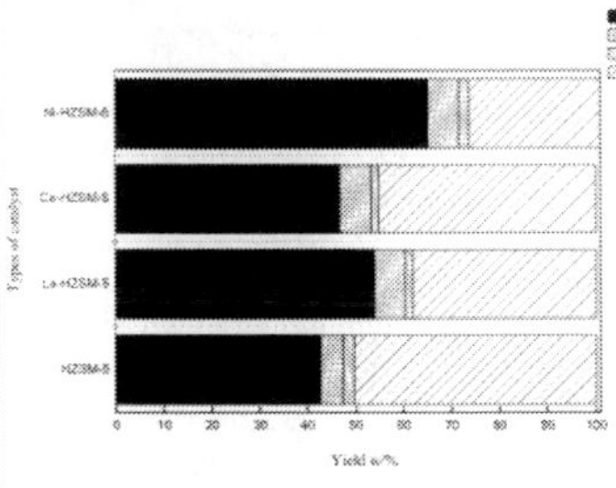

Fig. Products distribution on different catalysts
(At 450℃, feed rate=1.5 ml·min^{-1}, WHSV =4 h^{-1})

- Organic liquid yield increased
- Coking yield decreased
- Gas yield decreased

Part 3 Effect of mdified catalysts on conversions

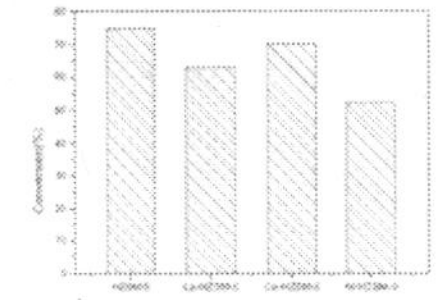

Fig. Total conversions of model components

- Total conversions decreased

- Single conversions decreased except EA over Ce-HZSM-5 and DAA

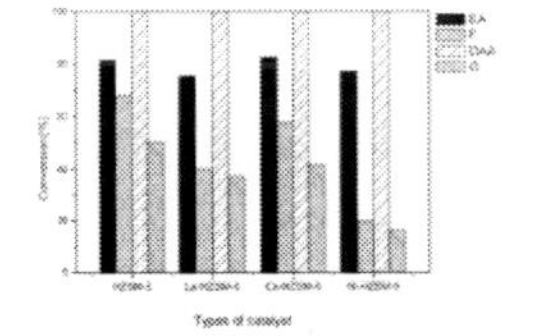

Fig. Single conversions of model component

Part 3 Deoxidizing products over HZSM-5

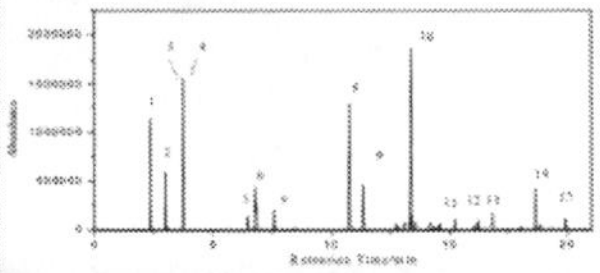

Fig. TIC of the major deoxidizing products of model components

Complete deoxidizing products

Partial deoxidizing products

Table Major deoxidizing products of model components

Number	[illegible]	Compound	[illegible]
1	2.343	Benzene	78
2	2.669	2-Methyl Furan	81
3	2.978	2-Vinyl Furan	94
4	3.737	Toluene	91
5	6.465	Ethyl benzene	91
6	6.628	M-Xylene	91
7	7.399	P-Xylene	91
8	10.795	Phenol	94
9	11.357	Benzofuran	118
10	13.374	O-Phenyl phenol	108
11	14.193	M-Cresol	108
12	15.250	2-Methyl benzofuran	131
13	16.817	Dihydrobenzofuran	120
14	18.043	Naphthalene	128
15	19.908	2-Ethyl benzofuran	131

Part 3 Effect of mdified catalysts on Deoxidizing degree

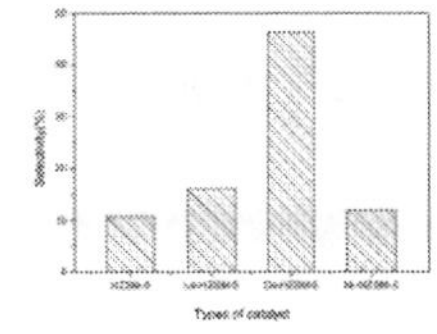

Fig. The selectivity of aromatic on catalysts

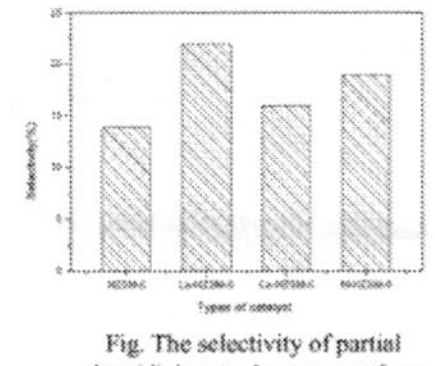

Fig. The selectivity of partial deoxidizing products on catalysts

- Selectivity of aromatic increased, Ce-HZSM-5 was the best.
- Selectivity of partial deoxididation increased, La-HZSM-5 was the best.

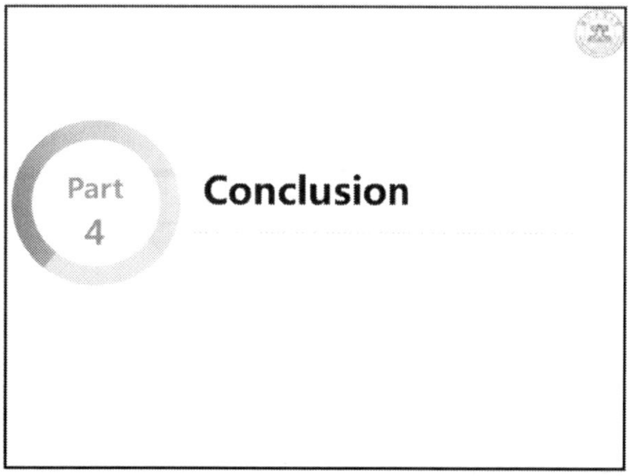
Part
4
Conclusion

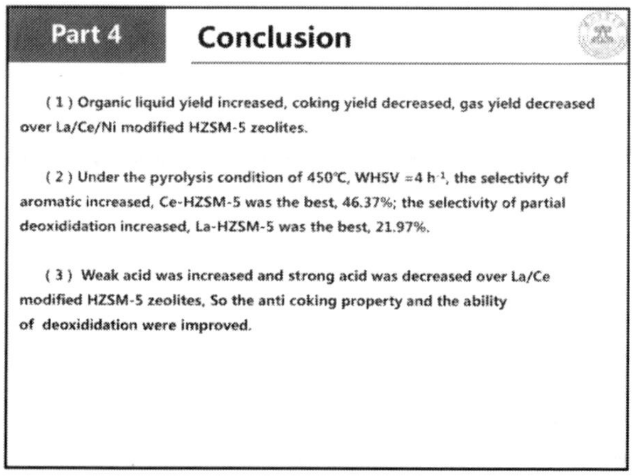
Part 4
Conclusion
(1) Organic liquid yield increased, coking yield decreased, gas yield decreased over La/Ce/Ni modified HZSM-5 zeolites.
(2) Under the pyrolysis condition of 450℃, WHSV =4 h⁻¹, the selectivity of aromatic increased, Ce-HZSM-5 was the best, 46.37%; the selectivity of partial deoxididation increased, La-HZSM-5 was the best, 21.97%.
(3) Weak acid was increased and strong acid was decreased over La/Ce modified HZSM-5 zeolites, So the anti coking property and the ability of deoxididation were improved.

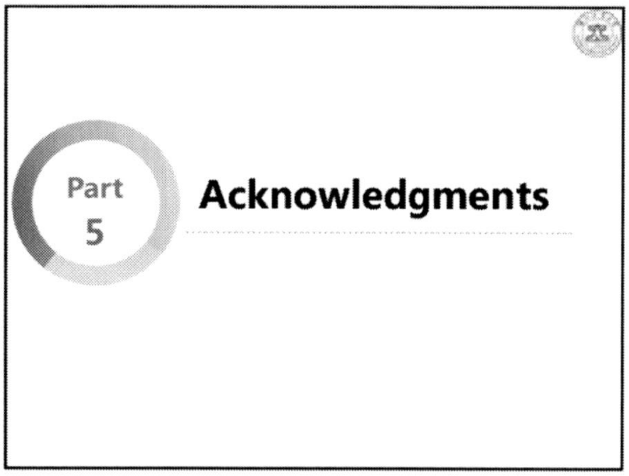
Part
5
Acknowledgments

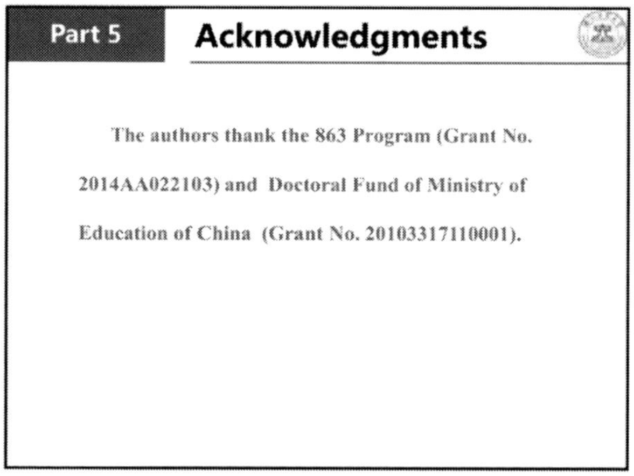
Part 5
Acknowledgments
The authors thank the 863 Program (Grant No. 2014AA022103) and Doctoral Fund of Ministry of Education of China (Grant No. 20103317110001).

Part 6 References

Part 6 References

[1] Eoin Butler; Ger Devlin; Dietrich Meier; Kevin McDonnell , A review of recent laboratory research and commercial developments in fast pyrolysis and upgrading [J]. Renewable and Sustainable Energy Reviews, 2011, 15(8): 4171-4186.

[2] Fengwen Yu; Longchao Gao; Weijin Wang; Guodong Zhang; Jianbing Ji. Bio-fuel production from the catalytic pyrolysis of soybean oil over Me-Al-MCM-41 (Me-La, Ni, or Fe) mesoporous materials[J]. Journal of Analytical and Applied Pyrolysis, 2013, 104:325-329

[3] W T Tsai, M K Lee, Y M Chang. Fast pyrolysis of rice husk: Product yields and compositions[J]. Bioresource Technology. 2007, 98(1): 22-28

[4] Sathish K. Tanneru, Philip H. Steele. Direct hydrocracking of oxidized bio-oil to hydrocarbons[J]. Fuel, 2015,154: 268～274

[5] A.V. Bridgwater. Review of fast pyrolysis of biomass and product upgrading[J]. Biomass and Bioenergy, 2012, 38:68-94.

[6] YANG Wen-yan, ZENG Yan, LUO Jia, TONG Dong-mei, QING Ren-wei, FAN Yong, HU Chang-wei. Production of bio-oil by direct and catalytic pyrolysis of Nannochloropsis sp. [J]. Journal of Fuel Chemistry and Technology，2011，39(9)：664～669.

[7] Supawan Vichaphund ; Duangdao Aht-ong;Viboon Sricharoenchaikul;Duangduen Atong. Production of aromatic compounds from catalytic fast pyrolysis of Jatropha residues using metal/HZSM-5 prepared by ion-exchange and impregnation methods [J]. Renewable Energy, 2015, 79:28-37.

[8] CHANG Sheng, ZHAO Zeng-li , ZHANG Wei, ZHENG An-qing, WU Wen-qiang, LI Hai-bin. Comparison of chemical composition and structure of different kinds of bio-oils[J]. Journal of Fuel Chemistry and Technology，2011，39(10)：746～753.

[9] Feiyan Gong, Zhi Yang, Chenggui Hong, Weiwei Huang, Shen Ning, Zhaoxia Zhang, Yong Xu, Quanxin Li. Selective conversion of bio-oil to light olefins: Controlling catalytic cracking for maximum olefins [J]. Bioresource Technology, 2011,102: 9247~9254.

[10] Ji Hongqiang , Zhang Qiang , Dong Xue, Shan Honghong. PERFORMANCE OF METHANOL AROMATIZATION ON NICKEL MODIFIED HZSM-5 CATALYSTS[J]. Petroleum Processing and Petrochemicals, 2015,46(2):58-63.

[11] Shanling Li;Suping Zhang;Zhanyuan Feng;Yongjie Yan. Coke formation in the catalytic cracking of bio-oil model compounds [J] Environmental Progress and Sustainable Energy , 2015,34(1): 240-247.

[12] S.L González-Cortés, S Rugmini, T Xiao, M.L.H. Green, S.M. Rodulfo-Baechler, F.E. Imbert. Deep hydrotreating of different feedstocks over a highly active Al_2O_3-supported NiMoW sulfide catalyst[J]. Applied Catalysis A:General, 2014, 475:270-281.

[13] Minghao Zhou, Yuan Wang, Yanbin Wang, Guomin Xiao. Catalytic conversion of guaiacol to alcohols for bio-oil upgrading[J]. Journal of Energy Chemistry, 2015,24 (4): 425-431.

[14] Wen Chen, Zhongyang Luo, Chunjiang Yu, Yi Yang, Guoxiang Li, Jixiang Zhang. Catalytic conversion of guaiacol in ethanol for bio-oil upgrading to stable oxygenated organics[J]. Fuel Processing Technology, 2014,126: 420-428.

[15] CHEN Jiao-jiao , CHEN Guan-yi , MA Wen-chao , MA Long-long, WANG Tie-jun, ZHANG Qi, LU Wei. Experimental study of aromatics production from catalytic cracking of bio-oil model compounds. Journal of Fuel Chemistry and Technology，2013，41(2)：183-188.

[16] Arakshita Majhi;Y.K. Sharma,Rajaram Bal, Babita Behera, Jagdish Kumar. Upgrading of bio-oils over PdO/Al2O3 catalyst and fractionation[J]. Fuel, 2013,107: 131-137.

浙江工業大學
ZHEJIANG UNIVERSITY OF TECHNOLOGY

THANKS!

464724 Process Modeling of Fluidized Bed Biomass-CO_2 Gasification Using Aspen Plus

Abstract:

Biomass can be converted to biofuels using various pathways which are broadly classified as biochemical or thermochemical. Gasification is a common thermochemical process that converts biomass to a gaseous product (syngas) which can be used for power generation, chemical production and as a fuel. The use of air, steam and oxygen as oxidizing agents in this process has been widely studied and reported. Improved CO concentrations have been observed when CO_2 is added as a co-reactive agent in the gasification of biomass.

In this work, a comprehensive model simulating the biomass gasification process using CO_2 in a fluidized-bed gasifier is developed using ASPEN Plus. The model utilizes additional reactors and mechanisms simulating various steps in the gasification process to predict the effect of operating parameters such as temperature and CO_2 flowrate on biomass conversion and syngas composition. A sensitivity analysis of product gas composition and a comprehensive sustainability assessment is conducted to evaluate the economic, environmental, and social performance of the biomass-CO_2 gasification process.

The results from this model are compared to published gasification experimental data obtained on a bench scale fluidized bed gasifier.

Copyright © American Institute of Chemical Engineers. All rights reserved.

451759 Simulation, Heat Integration and Rectisol-Based Decarbonisation for the Production of Synthetic Natural Gas from Biomass Gasification and Landfill Gases

Abstract:

The outline of this research is to develop a systematic heat-integrated gas purification process, followed by methanation for the production of synthetic natural gas or substitute natural gas (SNG). SNG production from biomass is currently considered as an alternative to reduce dependency on high-priced natural gas import and an opportunity to reduce greenhouse gases by applying CO_2 Capture and Sequestration (CCS) technologies. Biomass is thermally converted (e.g. gasification) to produce CO_2, CO, H_2O and H_2 gases that are cleaned to remove CO_2 before being converted to SNG (methane) via a methanation process.

The focus of this project is to develop a comprehensive simulation framework of heat-integrated decarbonized process suitable for the removal of contaminants from biomass-based produced gases to meet SNG production specifications.

In this study, a detailed heat-integrated steady-state simulation framework for biomass thermal conversion, water gas shift (WGS) reaction, gas purification by Rectisol process and SNG production via methanation process has been developed. The main objective was to explore heat recovery opportunities, cleaner operation by reducing contaminants and optimised selectivity with minimal cost. The simulation framework was constructed using Aspen HYSYS (V.2006) process simulator, which is capable to predict the optimised operating conditions of gasification, methanation and Rectisol process at which containments and solvent loss are minimized. Optimization software (SPRINT) is used for data extraction and analysis of heat integration opportunities across the entire process model.

In this project, a biomass (straw) was initially pyrolyzed to mainly produce a gas stream consisting of CO_2, CO, H_2O and H_2 along with tar and char. These streams were then mixed with landfill gases that comprise of methane (CH_4) (40-60%), CO_2 and H_2S as major components that also considered as major greenhouse gases (GHG). Furthermore, these gases are contaminated with sulphur, nitrogen, oxygen, siloxanes, water, halogenated compounds and many other chemicals which require complex gas clean-up process. This mixture was gasified to produce a gas stream consisting mainly of CO_2, CO, H_2O, H_2 and H_2S. The gas stream was then fed to a water gas shift reactor to increase H_2/CO ratio to 3. This gas mixture was cleaned up by a Rectisol process to remove CO_2, H_2S and other containments. The cleaned gas (CO, H_2O, H_2) is finally fed to a methanation process to convert it virtually to pure methane (CH_4), which is the desired final product (i.e. SNG).

The Rectisol process uses refrigerated methanol as a solvent, where 99% removal of CO_2 and H_2S that are particularly present in landfill gases and product gas from biomass gasification was achieved. Two-level refrigeration cycle was developed to maintain the operating temperature of the overall gas purification process. The methanation route was selected for the production of SNG. Methanation is an exothermic reaction and hence, a comprehensive sensitivity analysis was carried out to optimize the process by effective removal of heat of reaction and to achieve maximum methane (CH_4) formation. This analysis showed that four reactors were used to achieve maximum conversion CO to CH_4 which corresponded to CO conversion of 97.8%. After an in-depth heat integration study of the overall process model, 2.5 MW of steam at 320 °C was generated and 0.45 MW process heat recovery was achieved, which resulted in energy savings.

In summary, this project establishes process simulation framework and heat-integrated design of biomass gasification, gas purification and methanation technology. Extensive heat integration strategies have been established based on the following criteria; maximum heat recovery, cleaner and economical production of SNG. The research focused on determining optimal operating conditions for WGS reactor to maintain H_2/CO ratio of 3, for Rectisol Process to maintain 99% removal of CO_2 and H_2S and for methanation process to maximise the conversion rate to SNG. The following conclusions have been drawn from this research:

1. The overall efficiency of SNG production depends significantly on the energy content of the biomass feedstock.
2. Optimizing the refrigeration system for the Rectisol Process plays a vital role in minimizing the total energy requirement of the whole process.
3. Environmental and sustainability factors should always be taken into account prior to selecting the energy conversion process.
4. To meet the SNG specification regarding composition, relative density, calorific value and Wobbe Index; the methanation process must be optimised for maximum conversion of CO to CH_4.
5. Systematic heat integration of the overall process is essential to reduce the utility requirements and provides opportunity for heat and power generation.
6. The cost of electricity from steam generation and cost of carbon capture were estimated to be a function of biomass feedstock cost. It is concluded from this study that biomass feedstock cost is absolutely critical in determining the overall operating cost.

Copyright © American Institute of Chemical Engineers. All rights reserved.

461719 Production of Synthetic Gas through Agricultural Waste Using Fixed BED Gasifier

Abstract:

Gasification is a process that converts organic fuels or fossil fuels (carbonaceous materials) into synthetic gas. It is referred as a producer gas from agricultural waste using gasifier. Biomass can be obtained from living organisms like wood, waste materials and agricultural waste were to be gasified and convert into synthetic gas by gasification process. The results and experiments of solid waste by gasifier is operating based on stoichiometric calculations referred as three different types of agricultural waste like cotton crop waste, tree barks and tree leafs. The material with higher content of the moisture was tree barks waste that was noted as 13.3% whereas cotton crop waste had 12.23% of moisture content and tree leafs waste had 12.81% moisture content. The study showed that cotton crop waste had almost same production of synthetic gas as of tree barks but the productivity was achieved higher in tree barks which had greater quantity of fixed carbon which was noted as 14.13%. The tree leafs also had higher amount of fixed carbon was noted as 12.85% but the ash content in tree leaf was noted as 18.92 %. The composition of synthetic gas from cotton crop waste was 27.52% CO and 32.17% H_2. The composition of synthetic gas from Tree barks was 28.72% CO and 31.88% H_2. The composition of synthetic gas from tree leafs were 26.07% CO and 28.12% H_2. As examined the increasing fuelâ€™s demand and consumption in a wide country to complete and fulfill the requirement of fuel and energy to achieve the desired targets and goals in Pakistan through the production of alternative fuels with inexpensive way. An innovative process of synthetic gas from solid waste and agricultural biomass is gasification process to generate SYN gas that can be used as replace of natural gas, LPG, power generation and also used for the production of DME (dimethyl ether) to run the mechanical vehicles in the great performance where compared to other fuel resources.

Copyright © American Institute of Chemical Engineers. All rights reserved.

470455 Population Ensemble Modeling of Cellulose Dissolution: Insights for Efficient Biomass Processing

Abstract:

A major barrier to the efficient utilization of biomass is the recalcitrance to dissolution of semicrystalline cellulose. The present study addresses the kinetics of swelling and dissolution of cellulose particles in conditions emulating large-scale processing where the particles exhibit a distribution of size and degree of crystallinity. To this end, we have developed a model in which the behavior of a population of particles is obtained from an ensemble of individual cellulose particle dissolution models. The dissolution of individual solid cellulose particles is based on the relevant transport phenomena and kinetics and reveals decrystallization and disentanglement as two important and potentially rate-determinant steps in the process. The average value or the number distribution of any intra-particle property captured by the individual particle model can be determined by simulation of a sufficient number of individual particles such that ensemble averages are independent of the particle numbers and the computed particle distributions are acceptably smooth. Using this population model, various cellulose particle size and crystallinity distributions are analyzed for different dissolution parameters. The findings from this study would be useful for the rational design and optimization of pretreatment processes to reduce the particle size and degree of crystallinity, leading to enhanced biomass utilization.

Copyright © American Institute of Chemical Engineers. All rights reserved.

475618 Renewable Aromatic Hydrocarbon Production from Furfural By a Continuous Dual-Stage Hydrogenation-Cocracking Process

Abstract:

Aromatic hydrocarbon production from furfural by catalytic cracking over zeolite catalysts is significant for realizing biobased aromatics generation, because furfural can be produced from biobased sugars through different transformation routes such as liquid phase conversion and fast pyrolysis. However, due to the high unsaturation degree and high oxygen content, direct cracking of furfural faces low aromatic hydrocarbon yield and coking problems. Therefore, in this work, two improving methods namely hydrogenation pretreatment and methanol-assisted cocracking over HZSM-5 catalyst were combined to convert furfural into aromatic hydrocarbons, and the influence of hydrogenation catalysts was investigated. It was found that in the hydrogenation stage, although the furfural was nearly completely converted over $10Cu/SiO_2$, the furan ring was not efficiently hydrogenated. $5Ni\text{-}5Cu/SiO_2$ and $10Ni/SiO_2$ achieved the further hydrogenation of some furan rings, but over $10Ni/SiO_2$ the primary conversion of furfural was lower and more saturated gaseous hydrocarbons were generated. The different hydrogenation behaviors of furfural affected the subsequent cracking processes. The hydrogenation-cocracking of furfural and methanol over $5Ni\text{-}5Cu/SiO_2$ generated the highest oil phase yield of 26.4%, which was mainly composed of mono-aromatic hydrocarbons such as benzene, toluene, xylenes and trimethylbenzenes. However, because of the insufficient hydrogenation of furan ring over $10Cu/SiO_2$ and the low conversion of aldehyde group as well as some excessive hydrodeoxygenation over $10Ni/SiO_2$ in the hydrogenation stage, the yield and quality of the oil phases in the corresponding following cracking stage both decreased. The comparison of hydrogenation-cocracking process using the optimum hydrogenation catalyst $5Ni\text{-}5Cu/SiO_2$ with other traditional cracking processes including single-stage cocracking of furfural and methanol, hydrogenation-cracking of pure furfural using $5Ni\text{-}5Cu/SiO_2$ as the hydrogenation catalyst and direct cracking of furfural indicated that the hydrogenation-cocracking process was the most favorable technique for the generation of liquid aromatic hydrocarbons and the maintaining of catalyst activity. Finally, based on the experimental results, the hydrogenation-cocracking mechanism of furfural and methanol was proposed to explain the different hydrogenation behaviors of furfural over three catalysts and also the subsequent different cocracking performances.

Copyright © American Institute of Chemical Engineers. All rights reserved.

481269 Biomass Anaeorobic Convertion Technology Development

Abstract:

Base on the demands for environmental protection, energy security and modern agriculture development, with drawing on experiences of vigorously developing anaerobic biomass conversion from developed countries, innovative research on high efficient anaerobic conversion technologies has been conducted by Biomass Bio-chemical Conversion Laboratory, Chinese Academy of Sciences Guangzhou Institute of Energy Conversion.

In recent years, our Laboratory has undertaken more than 30 projects funded by the National Science & Technology Program, National High Technology Research and Development Program, National Natural Science Foundation, National Science & Technology Pillar Program, Provincial and Ministerial Science & Technology Programs and also programs supported by Chinese Academy of Sciences and corporations, focusing on bio-chemical pretreatments of lignocellulosic materials, nutrient balancing regulation base on complementary characteristics of substrates, high concentration co-digestion of multi-feedstock, two-stage continuous desulfuration, treatments of industrial waste (Cassava starch Dreg, Molasses Alcohol Wastewater etc.). Technologies and reactors for breeding/cultivation/pretreatment/anaerobic digestion of energy grassesï¼Œfermentable organics extraction from municipal solid wastes with multiple impurities, two-phase& three-stage anaerobic digestion, combined oil-gas-fertilizer production from food waste have been developed base on our research; integrated technological system for high-efficient, continuous and stable production of biogas from organic wastes has been established. Maximum organic loading rates of 8-10 kgVS/(m^3Â·d), highest solid concentration of 20% as well as stable operation can be achieved within the system; the best performance results of 75% methane content, 89% volatile solid removal rate, 70% energy conversion rate and 45% lignin removal rate have be obtained; the biogas production cost in this system can be lower than 0.5 yuan/m^3. The international leading position of this achievement has been appraised based on the aggregate technology and economic indicators, and it has won the First Award of the Science and Technology Progress of Guangdong Province and Third Award of Energy Science and Technology Progress.
The aim of this report is to systematically introduce our achievement, and we are looking forward to cooperation on relevant research and application with domestic and foreign organizations. This work is funded by national high-tech development programs(No.KFJ-EW-STS-138, KGZD-EW-304-1, 2012AA101802 etc..)

Copyright © American Institute of Chemical Engineers. All rights reserved.

474806 Oxidative Desulfurization of DBT with H_2O_2 Catalysed By TiO_2/Porous Glass

Abstract:

Oxidative desulfurization of DBT with H_2O_2 catalysed by TiO_2/porous glass

Chun Shen[†], Yujun Wang*[,‡], Jianhong Xu[‡], Tianwei Tan[†], Guangsheng Luo*[,‡]

[†]Beijing Key Laboratory of Bioprocess, College of Life Science and Technology, Beijing University of Chemical Technology, Beijing 100029, China

[‡]The State Key Laboratory of Chemical Engineering, Department of Chemical Engineering, Tsinghua University, Beijing 100084, China

Corresponding author: Tel: 86-010-62788568, Fax: 86-010-62788568

Email address: wangyujun@tsinghua.edu.cn and gsluo@tsinghua.edu.cn

Abstract

Aimed at ultra-deep oxidative desulfurization (ODS) of dibenzothiophene (DBT) and 4,6-dimethyldibenzothiophene (4,6-DMDBT) to control air pollution, we specially designed and prepared porous glass supported with TiO_2 nanoparticles acting as an amphiphilic catalyst. Hydrogen peroxide which is considered as the "green" oxidant was used, and for the extreme liquid–liquid phase ratio (usually larger than 1500) reaction system, the pore volume of 0.19 mL/g of the catalyst provides enough space for storage of hydrogen peroxide. The as-prepared catalyst offers a high interfacial surface area of 116.9 m^2/g and enhances the reaction by facilitating the mass transfer. The mono-dispersed TiO_2 exhibited good crystallinity. The mean diameter varied from 2.1 to 7.8 nm with the loading amount increased from 1.27 wt.% to 9.85 wt.%. The catalyst showed high activity and good stability for producing ultra-clean fuels: 100% conversion was obtained within 2 min and the conversion just decreased from 100.0±1.0 % to 94.3±0.6 % after 5 cycles. Overall, this new reusable catalyst provided an alternative for highly efficient ultra-deep desulfurization in a green way.

Copyright © American Institute of Chemical Engineers. All rights reserved.

475573 Hydrolysis of corncob using a modified carbon-based solid acid catalyst

Abstract:

Lignocellulosic biomass is a very important renewable resource that could be used to produce biofuels and various platform chemicals. The conversion of biomass to reducing sugar is an essential intermediate step that can impact the economic benefits of the producing process. The modified carbon-based solid acid is a novel catalyst used in the hydrolysis process. A catalyst material was prepared through the incomplete carbonization and sulfonation of the mixture of glucose and *p*-toluenesulfonic acid, followed by oxidation using 30% H_2O_2 to promote the acidity. The hydrolysis efficiency of the solid acid was significantly increased. The effects of the carbonization temperature, time, and the glucose / *p*-toluenesulfonic acid ratio on catalytic efficiency were investigated. Under the appropriate preparation condition, the catalyst could hydrolyze corncob to produce 78% xylose and 34% glucose. The activity of the catalyst slightly declined after five-cycles application.

Copyright © American Institute of Chemical Engineers. All rights reserved.

474464 Simultaneous Catalytic Conversion of Bio-Oil and Bio-Char for Hydrogen Production

Abstract:

Based on the integrated processes for hydrogen production from biomass, the simultaneous catalytic conversion of bio-oil and bio-char for hydrogen production was investigated in this paper. The potential hydrogen (H_2+CO) yield was increased from 45.70 to 82.19 mmol/g bio-oil with the presence of metals in the bio-char. The effects of metals (K, Ca, Mg, Fe, Zn, and Al) in bio-char on the catalytic conversion of bio-char and bio-oil were also investigated. The catalytic activities of the metals for the conversion of bio-char and bio-oil were followed by the sequence K>Ca>Mg>Fe>Zn>Al. Meanwhile, the alkali and alkaline earth metals (AAEM), including K, Ca, and Mg, were found to play the dominating roles among the metals in bio-char, and the catalytic effect of K was the most obvious. The potential hydrogen yield and tar yields were 143.50 mmol/g bio-oil and 4.96 mg/g bio-oil when adding the metal of 5 wt.% K.

The effects of metals on the biochar and biomass pyrolytic vapor (bio-oil) were investigated. Al had an inhibition on the catalytic activity of biochar, as a result, the soot yield was increased. Naphthalene was the primary tar in reaction temperature was 900 °C, also had a thermal stable in 900 °C. Fe was deactivated by coke and sintering easily.

The integrated process for hydrogen production from biomass was developed

Basing on the gas-solid simultaneous gasification process of the integrated processes of biomass, the effects of metals (K, Ca, Mg, Fe, Zn, and Al) in biochar on the catalytic gasification and catalytic reforming processes were investigated in this paper.

The catalytic reforming of biomass pyrolytic vapor (MPV) over biochar in a fixed bed was investigated. The carbon conversion and potential hydrogen yield of biomass pyrolytic vapor were increased respectively from 68.46% to 86.88% and 42.16% to 82.76% in the presence of biochar. The catalytic reforming activities of different metals were found with the following the sequence: K>Ca>Mg>Fe>Zn>Al. Meanwhile, the alkali and alkaline earth metals (AAEM) like K, Ca, and Mg were found to play the dominating roles in the conversion of MPV, and especially, the catalytic effect of K was the most obvious among the metals in bio-char. The carbon conversion and potential hydrogen yield of pyrolytic vapor were increased by increasing temperature and ratio of steam to pyrolytic vapor.

The effects of metals on the biochar catalytic gasification were also investigated. The catalytic activities of metals on biochar catalytic gasification were in the sequences: K>Ca>Mg>Zn>Fe>Al. AAEM (K, Ca and Mg) and Fe can promote water-gas shift reaction, while Al had an inhibition effect on water-gas shift reaction. The increasing of temperature and steam were benefit for biochar catalytic gasification. When the carbon conversion was more than 90% (S = 20 g/h and T =900 °C), biochar and demineralized biochar need 40 min and 115min respectively.

The effects of metals on the biochar and biomass pyrolytic vapor (bio-oil) simultaneous catalytic conversion were investigated. The potential hydrogen (H_2+CO) yield was increased from 45.70 to 82.19 mmol/g bio-oil with the presence of metals in the biochar. And the tar yield was decreased from 24.88 to 13.84 mg/g bio-oil. The catalytic conversion of biochar and bio-oil were followed by the sequence K>Ca>Mg>Fe>Zn>Al. Meanwhile, the potential of hydrogen and tar yields were 143.50 mmol/g bio-oil and 4.96 mg/g bio-oil when adding the metal of K. The performance of AAEM (K, Ca and Mg) was comparable with Ni in catalytic reforming bio-oil. AAEM can promote water-gas shift and methane reforming reactions, also promote the elimination of soot. Al had an inhibition on the catalytic activity of biochar, as a result, the soot yield was increased. Naphthalene was the primary tar in reaction temperature was 900 °C, also had a thermal stable in 900 °C. Fe was deactivated by coke and sintering easily.

Copyright © American Institute of Chemical Engineers. All rights reserved.

465897 The Pretreatment of Pennisetum Hybrid with Composite Solid Base Catalyst

Abstract:

The Pretreatment of Pennisetum hybrid with Composite Solid Base Catalyst

Zhuang Xinshu, Tan Xuesong, Yu Qiang, Qi Wei, Wang Wen, Wang Qiong, Zhou Gui Xiong,

Yuan Zhenhong*

(Key Laboratory of Renewable Energy, Guangzhou Institute of Energy Conversion, Chinese Academy of Sciences, Guangzhou 510640, China)

Abstract: Pretreatment is a critical step for the downstream enzymatic hydrolysis to producing cellulosic ethanol. Solid base catalyst can be a potential way for lignin removing, xylose releasing and phenols producing. In this paper the effects of the reaction conditions on enzymatic digestibility and total sugar yield of Pennisetum hybrid pretreated in low temperature aqueous environment with composite solid alkaliâ€"CaO/MgO as the catalyst are investigated. The results indicated that 20% catalyst content, 110â„ƒ,80min and 1:20 ratio of solid are the optimal pretreatment conditions, the lignin removal rate is up to 54.18%.The enzymatic digestibility increases from 41.23% to 81.04% with the enzyme loading of 40FPU/g after72h, substrates can be fully hydrolyzed. After pretreatment, the compact structure of the feedstock is damaged, more fibers exposing, the chemical groups such as hydrogen bond in or between are destroyed.

Keywords: Composite Solid Base; Pennisetum hybrid; Enzymolysis; Pretreatment

This work is financially supported by the National Natural Science Foundation of China (51476179) and the Sino-Thai project (51561145015).

Corresponding author: Yuan Zhenhong, 86-20-87057735; yuanzh@ms.giec.ac.cn.

Copyright © American Institute of Chemical Engineers. All rights reserved.

474519 Conversion of Cheese Whey to Lactobionic Acid

Abstract:

Lactobionic acid (LBA) is a versatile, high-value, low-volume chemical with numerous applications in the food, pharmaceutical, cosmetic, and chemical industries. Microbial production of LBA has the advantages of high selectivity and the ability to use cheese whey, a waste stream from the dairy industry, which is rich in lactose. However, microbial LBA production has not yet been developed as successfully as other organic acids, despite attempts being made to use wild type bacteria, fungi and yeast strains. There is no commercial implementation of microbial LBA production technologies due to the following factors: 1) The most robust strains are pathogenic, which prevents their industrial implementation. 2) The enzyme (cellobiose dehydrogenase or lactose dehydrogenase) which oxidizes lactose to LBA is produced only in the stationary phase. The highest level production of LBA was only achievable under resting cell conditions, which are not usually used in industrial fermentations. 3) The microorganism consumed LBA at the later stage of the fermentation. We report engineering a Neurospora crass strain for cellobionic acid production from cheese whey and solving the problems associated with LBA fermentation using metabolic engineering efforts

Copyright © American Institute of Chemical Engineers. All rights reserved.

468385 Predictive Model and Bioconversion of Mixed Feedstocks

Abstract:

Currently, biorefineries are utilizing single feedstock inputs but in order to reduce feedstock dependency and the related risks, researchers are exploring the option of mixing multiple feedstocks. To understand the implication of mixing feedstocks on sugar yield, we relied on statistical design of experiments to build a predictive model for sugar yield based on varying feedstocks fractions, pretreatment types, and pretreatment reaction parameters. We performed pretreatments at low biomass loading, but for a biorefinery to be economically viable biomass deconstruction has to be performed at higher solid loading. Also, lignocellulosic sugars should be fermentable to produce high energy density molecules such as hydrocarbons.

In this study, three biomass feedstocks: corn stover, switchgrass, and energy cane were chosen based on the least cost formulation model developed by Idaho National Laboratory. These single and mixed feedstocks were pretreated with dilute alkali (DL), dilute acid (DA), and ionic liquid (IL) at severities predetermined by the statistical experimental design and were subsequently, enzymatically hydrolyzed with CTec2® at a loading of 10 mg protein/g glucan. The resulting sugar yields from all treatments, which varied from 25 to 100% w/w untreated biomass, were then fed into SAS JMP® to develop a predictive model that generated a continuous envelope of optimal feedstock mixtures. In order to achieve 90% sugar yield, we have to perform IL pretreatment at 140°C for 120 mins with 12%, 13%, and 75% (w/w) corn stover, switchgrass, and energy cane. More predictions of such kind were generated from this model that can potentially overcome the dependence on a single feedstock while maintaining the overall product yields.

These feedstock/treatment combinations, selected from the model, were tested at higher biomass loading to study the impact on downstream fermentation process to produce ethanol and hydrocarbon, such as bisabolene. At higher biomass loading, mixed feedstock conversion to ethanol performed very similar to single feedstock conversion to ethanol. Rheological characterization was performed on high solid pretreated biomass mixtures to better understand mixing, material handling, and mass transfer issues. This information will be applied in techno-economic analysis of this end-to-end study of deconstruction of optimal biomass mixtures at high solid loading for ethanol and hydrocarbon production.

Copyright © American Institute of Chemical Engineers. All rights reserved.

459583 Two-Phase Kinetics of Enzymatic Depolymerization of Hemicellulose to Soluble Sugars for Liquid Biofuel Production

Abstract:

Hemicellulose is the second most abundant polysaccharide in nature, available not only in its pure form but also as a constituent of all lignocelluloses. The conversion of hemicelluoses to soluble sugars in parallel to its cellulosic counterpart would make liquid biofuels competitive in the market. Of the three key conversion steps (pretreatment of biomass, enzymatic hydrolysis of pretreated biomass to soluble sugars, ethanol from sugars by fermentation), hydrolysis is the slowest and the rate limiting step [1]. Adsorption of the liquid phase enzymes (endo- and exo-xylanases) to the insoluble solid substrate (xylan) is the first step in hemicellulose hydrolysis, followed by enzyme-catalyzed xylan depolymerization reactions in the solid and liquid phases. This work uses a coupled theoretical and experimental framework [2] to quantify the kinetics of product-inhibited solid and liquid phase depolymerization reactions that produce xylose and other reducing (soluble) sugars. We experimentally determine the type of adsorption of the enzyme on the solid substrate, the type of product inhibition in the solid and liquid phases, the transition of two-phase hydrolysis to single in batch reactors. Model-experiment comparisons allow us to quantify the xylobiose inhibition constants in the two phases. Our experiments show that the hydrolysis system transitions from a two-phase (solid-liquid) to a single (liquid) phase process at 11 hours. The enzymatic hydrolysis in both liquid and solid phases is found to be non-competitively inhibited by the product (xylose), with the inhibition being more potent in the liquid phase. Our multi-step reaction kinetic model is simulated using the liquid and solid phase kinetic data and validated with our experimentally measured temporal dynamics of xylose and reducing sugar yields in the liquid phase, and of the solid phase hemicellulose concentrations. We show that the product inhibition in the solid phase increases with the substrate loading, thus lowering the yield of reducing sugar, while the liquid phase product inhibition decreases with substrate loading, thus increasing the xylose yield. Our substrate inhibition studies allow us to recommend 5 mg/ml as the optimum xylan loading, above which substrate inhibition is observed to set in.

References:

[1] Chakraborty, S., Raju, S., and Pal, R., "A multiscale three-zone reactive mixing model for engineering a scale separation in enzymatic hydrolysis of cellulose," *Bioresource Technology*, **173, pp 140-147** (2014)

[2] Dutta, S.K., and Chakraborty, S., "Kinetic analysis of two-phase enzymatic hydrolysis of hemicellulose of xylan type," *Bioresource Technology*, **198, pp 642-650** (2015)

Copyright © American Institute of Chemical Engineers. All rights reserved.

474465 Selective Fermentation of Six Carbons Sugars from Hot-Water Hardwood Extract Hydrolysates

Abstract:

Hot water wood extract hydrolysate contains a mixture of sugars: xylose, glucose, galactose, mannose, arabinose and rhamnose. This sugar mixture presents a challenge in fermentation to ethanol and other chemicals / biofuel. The fermentation efficiency is especially low for xylose. When genetically modified *E. coli* fbr5 was employed to ferment the sugar mixture to ethanol, glucose was the first sugar consumed, and then galactose, arabinose, mannose and rhamnose, xylose was the last to be consumed. While *E. coli* fbr 5 was genetically engineered to ferment Xylose, this outcome is disappointing. Most microorganisms prefer glucose, *E coli* is no exception. Xylose being the dominant sugar species in hardwood hot-water extract hydrolysate and the least preferred sugar for ethanol production present a challenge to the biomass energy industry development. On the other hand, xylose can be employed to produce xylitol, furfural, and a lot more products efficiently. The value of xylose is higher than glucose in the market, and yet it is limiting the efficiency in biomass conversion. Hot water wood extract hydrolysate was subject to fermentation with *S. cerevisiae* and xylose was left unconsumed. Xylose can be recovered from the fermentation broth, while ethanol was produced from glucose.

Copyright © American Institute of Chemical Engineers. All rights reserved.

465914 Effect Temperature and Flow Rate on Lhw Pretreatment SB: Kinetic Study

Abstract:

Assuming xylan hydrolysis to xylose in sugarcane bagasse (SB) by liquid hot water (LHW) and xylose further degradation to furfural were two consecutive first-order reactions, a kinetic model was developed. SB was pretreated by LHW at different temperatures and flow rates, and xylan hydrolysis behavior in SB pretreatment was investigated using the developed kinetic model. Results showed the kinetic model could fit the data well (R^2>0.85). k_S (the rate constant of xylose production) and k_X (the rate constant of xylose degradation) at different temperatures was in accordance with Arrhenius equation. Activation energy (E_a) for xylose production and xylose degradation was 50.39 and 104.66 kJ/mol, respectively, which indicated that xylose degradation was relatively more difficult that xylose production during LHW hydrolysis. The relationship between k_S (or k_X) and flow rate was almost linear. The application of the model in xylan hydrolysis kinetics was successful.

Copyright © American Institute of Chemical Engineers. All rights reserved.

474466 R-HB Production from Glucose By B. Cepacia

Abstract:

Polyhydroxylbutyrate (PHB) is one of the important Polyhydroxyalkanoates (PHAs) that are a promising alternative to conventional, fossil fuel derived plastics. Plastic is placing a burden on the environment by polluting ecosystems, overwhelming landfills, using finite oil resources, and contributing to greenhouse gas emissions. PHAs are polyesters created by microorganisms intracellularly under limited nutrient growth conditions as carbon and energy storage compounds. PHAs are not yet widely accepted and produced because of their high production costs, specifically the carbon source and recovery method, compared to those of petroleum based plastics. In addition, present technology in recovery PHA is not environmentally friendly, leading to whether PHA is a sustainable bioproduct. This presentation shows some of our research efforts in utilizing renewable carbon source. Glucose was employed to produce droxybutyric acid. Butyric acid can be polymerized to produce PHB. The production of butyric acid and polymerization to PHB enables the desired biodegradable plastics to be recyclable.

Copyright © American Institute of Chemical Engineers. All rights reserved.

467000 Process Strategies for High Titers of Lipid Production By Oleaginous Yeasts in Undetoxified Hydrolyzates of Lignocellulosic Biomass

Abstract:

Oleaginous yeasts can accumulate up to 70% of cell biomass as lipid, predominantly as triacylglycerols. Yeast lipid fatty acid profiles have been reported to be similar to that of vegetable oils and consist primarily of oleic, palmitic, stearic and linoleic acids. This capability provides the opportunity to use yeasts to produce bio-based fuels and chemicals from agricultural residues and bioenergy crops, such as corn stover and switchgrass, respectively. Successful cultivation of these yeasts on inhibitory lignocellulosic hydrolyzates is uncertain because little has been published regarding their application to bioconversion of plant biomass. Several novel yeasts producing promising lipid concentrations (>10 g/L) from enzyme hydrolyzates of acid and base pre-treated lignocellulosic biomass were discovered in a prior screening of the ARS Culture Collection (NCAUR, Peoria, IL). Three of the top strains have not previously been reported for the bioconversion of lignocellulose to lipids. Process strategies for maximizing lipid production in one of the more benign hydrolyzates, AFEX-pretreated corn stover, and one of the more toxic hydrolyzates, dilute acid-pretreated switchgrass, respectively, were employed. Hydrolyzates were prepared at up to 20% solids loading to provide over 100 g/L sugars (at ~56:39:5 mass ratio glucose:xylose:arabinose). A two stage process boosting the molar C:N ratio from 60 to well above 400 in undetoxified switchgrass hydrolyzate was optimized with respect to nitrogen source, C:N, carbon loading and pH. Using this process three strains were able to consume acetic acid and nearly all available sugars to accumulate 50-65% of cell biomass as lipid (w/w), to produce 25-30 g/L lipid at 0.12-0.22 g/L/h and 0.13-0.15 g/g (or 39-45% of the theoretical yield) at pH 6 and 7, a performance unprecedented in lignocellulosic hydrolyzates. Depending upon plant biomass yields, we estimated that oleaginous yeasts have the capability of producing ~48 and 190 gal oil per acre from corn stover and switchgrass, respectively. As a frame of reference, ~48 gallons of oil per acre are produced from processing soybeans. The successful application of top-performing lipid-producing yeast in undetoxified lignocellulose hydrolyzates is expected to advance the economic feasibility of high quality biodiesel and jet fuels from renewable biomass, expanding the market potential for lignocellulose-derived fuels beyond ethanol for automobiles to the entire U.S. transportation market.

Copyright © American Institute of Chemical Engineers. All rights reserved.

471952 Evaluation of Alcohol-to-Jet (ATJ) Conversion Technology for Renewable Jet Fuel

Abstract:

Abstract Submission to Be Considered As An Oral Presentation at AIChE 2016 Annual Meeting in the Biochemical Conversion Processes in Forest/Plant Biomass Biorefineries Session

Evaluation of Alcohol-to-Jet (ATJ) Conversion Technology for Renewable Jet Fuel

Scott Geleynse, Manuel Garcia-Perez, Michael Wolcott, Xiao Zhang

Washington State University

The FAA is working to enable the U.S. use of one billion gallons per year of "drop-in" sustainable alternative jet fuels by 2018. Airlines and aerospace manufacturers around the world have also expressed continuously increasing interest in the replacement of jet fuel from fossil fuels to renewable fuels from biomass. Despite the considerable investment in the research of several different conversion strategies, establishing the successful commercial production of biorenewable jet fuel remains a challenge. In this presentation, we will discuss technical aspects of strategies to use alcohols as a platform molecule to generate jet fuel and other products. By using process modeling and techno-economic analysis, we aim to assess the feasibility and readiness of the technology for biorefinery implementation. We hope to provide insight into the various political and economic challenges faced in making and meeting renewable fuel goals.

Presenting author:

Scott Geleynse

Bioproducts Sciences and Engineering Laboratory

School of Chemical Engineering and Bioengineering, Washington State University
2710 Crimson Way, Richland, WA 99354

Contact: scott.geleynse@email.wsu.edu, x.zhang@wsu.edu

Copyright © American Institute of Chemical Engineers. All rights reserved.

473497 Kinetics of the Enzymatic Hydrolysis of Paper Pulp Fibers

Abstract:

Papermaking fibers, particularly waste fibers from recycling operations can be a valuable resource for bioenergy and products such as ethanol, bioplastics or biochemicals. The amount of lignin in such fibers is much lower than native wood or other plant based lignocellulosic resources eliminating the need for pretreatment steps. Moreover, waste fibers are readily available in the waste streams rejected from the papermaking operations to improve paper strength and runnability of the paper machines.

The present work develops a model for the kinetics of the enzymatic hydrolysis of cellulosic pulp fines based on the physico-chemical interactions. The cellulolytic enzymes adsorb to the outer surfaces of the fines after which they diffuse into the interior pores. The fine particles are assumed to be spherical, containing two distinct kinds of pores: one at the macro-scale and another at a micro-scale allowing slow progression of the enzymes into the solid cellulosic matrix. Enzymatic hydrolysis reactions are assumed to occur at the micro-scale within the cellulosic matrix. Convective mass transfer boundary layers reduce the enzyme transport across the external particle surfaces. The main utility of this model is in estimating the optimal reaction conditions and assist in scaling studies of these enzymatic hydrolysis processes.

Keywords: Enzymatic hydrolysis, cellulose, pulp fibers, fines, rejects, paper mill waste solids

Copyright © American Institute of Chemical Engineers. All rights reserved.

472678 Comparing Sugar Titers and Ethanol Yields from Fed Batch Fermentation of Celf and DA Pretreated Poplar Solids at High Glucan Loadings

Abstract:

Biological conversion of lignocellulosic biomass to liquid fuels reduces greenhouse gas emissions and addresses other economic and strategic challenges. However, costs are high due to high enzyme loadings and capital costs, particularly at glucan loadings over 10% needed to realize economically attractive ethanol concentrations. Because solids concentrations above about 10% cannot be readily mixed in batch fermentations, a fed-batch strategy was applied to establish maximum carbohydrate conversions from enzymatic hydrolysis of THF Co-Solvent Enhanced Lignocellulosic Fractionation (CELF) and dilute acid pretreated poplar wood. At 15, 18 and 20% glucan loadings, glucose concentrations of 63, 70, and 75g/l were achieved in only 2 days from CELF pretreated solids while concentrations from hydrolysis of dilute acid solids were only 45, 53, and 60g/l, respectively.

Copyright © American Institute of Chemical Engineers. All rights reserved.

458379 Co-Production of Bioethanol and Bio-Lactic Acid in a Biorefinery Concept: A Comprehensive Study

Abstract:

Biorefineries are seen as a key component for the transition toward a sustainable and low carbon bio-economy. In this research, alternative biorefineries annexed to a typical South African sugar mill are investigated to co-produce ethanol (EtOH), lactic acid (LA) and/or electricity, utilizing bagasse and a component of harvesting residues (brown leaves) as feedstock. Studied scenarios are EtOH production (scenario 1), EtOH form glucose and LA from xylose (scenario 2), EtOH form xylose and LA form glucose (scenario 3) and LA production (scenario 4), all of which are associated with some level of export electricity production. Aspen simulation of the scenarios were developed considering all supplementary units such as evaporation, water treatment, boiler and steam/power generation in addition to the process units. Economic evaluation, energy assessment and Life Cycle Analysis (LCA) of the developed simulations were carried out to determine the viability of each scenario. In order to supply the combined energy demands of both sugar mill and biorefinery, a 35 to 40% bypass of lignocellulose directly to the boiler section was required, to achieve integrated scenarios that were bio-energy self-sufficient, with no supplementation with fossil fuels. Based on the economic analysis, scenario 4 achieved the highest Internal Rate of Return (IRR) of 20.9%, while scenario 1 had the lowest IRR of 10.7%. Since EtOH, LA and enzyme prices strongly affect the profitability of biorefinery scenarios, sensitivity analysis was carried out, which identified scenario 2 as the most economically robust, having the least sensitivity to these price variations.

Copyright © American Institute of Chemical Engineers. All rights reserved.

474486 Thermophiles and Their Thermostable Enzymes in Biofuel Synthesis

Abstract:

Second-generation feedstock, especially nonfood lignocellulosic biomass, is a potential source for biofuel production. Cost-intensive physical, chemical, biological pretreatment operations and slow enzymatic hydrolysis carried out at â‰¤50Â°C make the overall process of lignocellulosic conversion into biofuels less economical than available fossil fuels. Thermophiles and their enzymes can play important roles in many kinds of bioprocessing including in conversion of non-food biomass into biofuels. The Homestake gold mine (8000 ft. deep, Lead, SD) offers a unique opportunity for direct exploration of the deep biosphere environment. Using soil/biofilm samples of deep biosphere of the Homestake Gold Mine, compost facility (Rapid City, SD, USA), and Hot Springs State Park (Thermopolis WY, USA), we have isolated several thermophilic cellulose- and xylan-degrading and -fermenting pure cultures belonging to the genera *Brevibacillus*, *Paenibacillus*, *Clostridium*, *Bacillus*, and *Geobacillus*. Unique characteristics of lignocellulose-deconstructing enzymes produced by mine and compost thermophiles include optimum temperatures of >70°C, pH ranges from 4 - 8, and high thermostability (e.g., at 60°C, 50% of cellulases and xylanases activities were retained in 35 and 23 days of incubation, respectively). Hot Springs State Park thermophiles can grow on various inexpensive regional carbon and energy sources (e.g., prairie cord grass and corn stover), and produce biohydrogen in a single step bioprocessing of biomass. This talk will describe the limitations in exiting lignocellulose conversion technologies and possible ways to overcome those limitations using thermophiles and their enzymes. The influence of high temperatures on various existing lignocellulose conversion processes and those that are under development, including separate hydrolysis and fermentation, simultaneous saccharification and fermentation, and extremophilic consolidated bioprocess will discussed. Integrated decentralized thermophilic biofuel production employing cellulolytic- and fermentative-thermophiles in a single step consolidated process will also be discussed.

Copyright © American Institute of Chemical Engineers. All rights reserved.

465690 Processing Carbonaceous Feedstock Using Ionic Liquids for Sustainable Carbon Engineering

Abstract:

Energy security is an important issue for the sustainable development of the global community. The United States faces major energy security challenges due to the high energy demand from residential, commercial, industrial, and transportation sectors, which spur the development of low energy and clean processes that produce fuels and chemicals from domestically available carbon-based feedstocks. Using lignocellulosic biomass to produce biofuels for transportation has emerged as a focus of interest in recent years. Biomass provides a renewable carbon-based source to reduce GHG emissions through the capture of CO_2 during photosynthesis, and decrease the reliance on petroleum for the production of both fuels and chemicals. However, biomass conversions to fuels and chemicals also suffer from its' own set of problems: the distributed nature of biomass resources, low energy density, high cost of transporting sufficient biomass to allow refineries to take advantage of economies of scale.

One promising alternative is to develop a hybrid feedstock system. This hybrid energy system by combining biomass with other carbonaceous feedstocks in a single conversion platform would reduce supply fluctuations, mitigate GHG emissions, and be more politically palatable. However, current research has mostly focused on thermochemical conversion of single type of feedstock at high temperature and elevated pressure, which requires high energy consumption. Developing low energy, environmental benign, feedstock agnostic technologies, is critically important to address concerns of energy security and process economics. Ionic Liquids (ILs) as a class of novel environmental benign solvents are receiving increased attentions. Many researches have focused on ILs' applications in lignocellulosic biomass dissolution for biofuels production, while the work on blending biomass with other feedstock sources to reduce the supply cost and improve the conversion efficiency is limited. This works looks into the feasibility of processing woody biomass mixture with other carbonaceous feedstocks using IL solvents for production of clean energy and high valuable materials. Various types of ILs and feedstocks, and process conditions are investigated. The IL processed carbonaceous materials are further characterized for their compositional, physiochemical and rheological properties changes to advance process optimization and development towards an economic feasible carbon engineering technology by employing wide range of regional feedstocks.

Copyright © American Institute of Chemical Engineers. All rights reserved.

469354 A New Lignin Based Polymer

Abstract:

Lignin, the most abundant renewable aromatic polymer on earth, is usually treated as a waste stream in the biorefinery industry. Developing new applications for lignin as valuable products or materials is a well-recognized challenge. Here, we present a template free method for synthesizing lignin-resorcinolâ€“formaldehyde porous gel with high lignin loading, by utilizing lignin with low molecular weight and less condensation. Up to 75% lignin loading was obtained by blending lignin.

Copyright © American Institute of Chemical Engineers. All rights reserved.

471770 Process Modeling and Techno-Economic Assessment of the Conversion of Biorefinery Lignin to Dicarboxylic Acids

Abstract:

The development of commercially viable, integrated biorefineries remains a difficult challenge. A key barrier is the lack of a strong value proposition for the lignin waste stream. Most approaches towards lignin valorization have focused on the depolymerization of lignin to phenolic compounds, but a class of chemicals â€"dicarboxylic acidsâ€" have largely been overlooked as a potential lignin product. Dicarboxylic acids are important and highly valuable industrial chemicals and intermediates used across the biopolymer, pharmaceutical, and food additives industries. Though the oxidative aromatic cleavage of lignin is a well-known reaction during the delignification of wood pulp, it is only recently that the selective production of dicarboxylic acids from biorefinery lignin has been demonstrated as a mild, environmentally friendly process. Development of this new pathway and its potential impact on biorefineries remains to be explored.

In this study, we will present several process models for the oxidative conversion of biorefinery lignin to dicarboxylic acid products, developed in Aspen Plus. We will evaluate and compare their technical and economic feasibilities, and present sensitivity analysis on key process parameters. Process and cost bottlenecks will be identified and discussed as potential areas for future research and improvement.

Copyright © American Institute of Chemical Engineers. All rights reserved.

472014 Integration of Renewable Jet Fuel Production with the Pulp Industry through Alcohol Conversion

Abstract:

Abstract Submission to Be Considered As An Oral Presentation at AIChE 2016 Annual Meeting in the Chemical Conversion Processes in Forest/Plant Biorefineries Session

Integration of Renewable Jet Fuel Production with the Pulp Industry through Alcohol Conversion

Scott Geleynse, Senthil Subramaniam, Xiao Zhang

Washington State University

Among increasing interest in the production of alternative fuels from biorenewable resources, considerable research has provided numerous routes to renewable jet fuels. Along with these new technologies, there is need for design concepts to integrate them with existing industries in order to enable new facilities and ease the transition toward the new renewable biorefinery industry. In this presentation, we explore one such idea in which the alcohols-to-jet conversion technology is implemented into an established kraft pulp mill. By utilizing a large portion of a pulp mill's processes, a biorefinery can be constructed on already established supply lines and facilities with significantly reduced capital investment. This may not only be a possibility to enable to generation of new renewable jet fuels on a large scale, but an option for pulp mills to benefit by reinventing their process and enter entirely new markets.

Presenting author:

Scott Geleynse

Bioproducts Sciences and Engineering Laboratory

School of Chemical Engineering and Bioengineering, Washington State University
2710 Crimson Way, Richland, WA 99354

Contact: scott.geleynse@email.wsu.edu, x.zhang@wsu.edu

Copyright © American Institute of Chemical Engineers. All rights reserved.

472054 Novel Solvent Deconstruction of Woody Biomass and Isolation of High Purity Lignin

Abstract:

Abstract Submission to Be Considered As An Oral Presentation at AIChE 2016 Annual Meeting in the Chemical Conversion Processes in Forest/Plant Biomass Biorefineries Session

Novel Solvent Deconstruction of Woody Biomass and Isolation of High Purity Lignin

Karissa Garcia, Ruoshui Ma, Scott Geleynse, Xiao Zhang
Washington State University

Lignin is the most abundant renewable aromatic polymer on territorial earth present in all high plants. Emerging interests in commercialization of lignocellulosic biofuel process injected new enthusiasm to lignin valorization. Due to the main objective of optimizing cellulosic fiber and/or carbohydrate recovery, most of current lignin isolated from either paper making or biorefineries are often difficult to purify and/or has been indiscriminately modified depending on the process conditions. It appears that one strategy has been overlooked is to devise a new method and/or identify a new solvent to extract lignin from plant biomass with a high purity and minimized structural heterogeneity.

Herein, we will report a new method of applying Deep Eutectic Solvents (DES) for extracting lignin from woody biomass with high yield and high purity. Four different DES were evaluated for treatment of hardwood and softwood. It was found that these DES treatments can selectively extract a significant amount of lignin from wood with high yields: 78% from hardwood and 58% from softwood. The extracted lignin has high purity (95%) with unique structural properties. The mechanism of DES cleavage of ether bonds between phenylpropane units was investigated.

The results from this study demonstrate that DES is a promising solvent for wood delignification and the production of a new source of lignin with promising potential applications.

Presenting author:

Karissa Garcia

School of Chemical Engineering and Bioengineering

Washington State University

Contact: karissa.garcia@wsu.edu, x.zhang@wsu.edu

Copyright © American Institute of Chemical Engineers. All rights reserved.

474861 A Model Algae Strain Extracellular Sugars: Genome Sequencing and Transcriptomic Analysis

Abstract:

Microalgae are viewed as a next generation biofuel platform. This perception is based upon rapid rates of growth, an ability to grow under a wide range of condition, and the production of a variety of different bioproducts, including lipids, proteins and carbohydrates from various microalgae species. A great deal of interest has centered on the production of lipids which could be converted into biodiesel, while the use of starch has been utilized to produce products such as ethanol in certain species. Microalgae have also been reported to produce extracellular carbohydrates, including complex polysaccharides as well as simple sugars such as glucose and maltose. Our laboratory has been studying the production of these simple sugars by a specific strain of *Chlorella* which was isolated from a symbiotic relationship with *Paramecium bursaria*. The *Chlorella* strain has been cultivated independently for several decades now, and continues to produce extracellular sugars under certain growth conditions. We have recently sequenced two strains of Chlorella species, and have completed RNA-Seq analysis of the expression profiles of this strain during growth conditions that produce sugars versus those that do not to compare the transcriptome and elucidate potential pathways toward extracellular sugar production. Results from these studies will be presented.

Copyright © American Institute of Chemical Engineers. All rights reserved.

461039 Exploiting Models to Assess and Guide Genetic Engineering of Microalgae Strains

Abstract:

Exploiting models to assess and guide genetic engineering of microalgae strains

Andrea Bernardi[1], Andrea Meneghesso[2], Giorgio Perin[2], Tomas Morosinotto[2], Fabrizio Bezzo*[1] [1]CAPE-Lab (Computer-Aided Process Engineering Laboratory) & PAR-Lab (Padova Algae Research Laboratory), Department of Industrial Engineering, University of Padova

via Marzolo 9, 35131 Padova, Italy [2]PAR-Lab (Padova Algae Research Laboratory), Department of Biology, University of Padova, via Ugo Bassi 58B, 35131, Padova, Italy *fabrizio.bezzo@unipd.it

Microalgae are long been recognized as one of the most promising alternatives for biofuel production. The main advantages of microalgae with respect to other possible feedstock are the high potential productivity and the absence of competition with traditional crops for arable land and clean water. However, this potential is still theoretical and algae production on large scale is not profitable yet. Several issues need to be addressed to reach this objective, ranging from algae cultivation and harvesting as well as products extraction [1]. In order to be economically and energetically sustainable microalgae cultivation must use the solar irradiance, which is variable on daily and seasonal basis. One of the main problems of microalgae cultivation in large scale is the low conversion efficiency of the solar energy. In fact, there is a big gap between the conversion efficiency obtained in laboratory and in large scale photobioreactors.

Microalgae have developed a series of mechanisms to modify their photosynthetic apparatus in order optimize the light utilization in a natural environment, such as variation in pigment composition and/or spatial organization of the photosynthetic apparatus [2]. In particular, two main regulatory mechanisms can be identified: photoregulation, often referred as Non Photochemical Quenching (NPQ), and photoacclimation. Photoregulation involves the activation of dissipative pathways to convert part of the light energy in thermal energy in order to prevent photo-induced damage (photoinhibition) and acts in a time scale from minutes to hours. Photoacclimation is a slower process acting in time scales of days and induced a variation on the pigment content of the microalgal cells as consequence of a variation in the light conditions. However, certain properties of microalgae, which have been positively selected in wild-type species through evolution, can cause a significantly loss of productivity and are therefore detrimental to large-scale cultivation as the growing conditions in industrial plants are very different from the natural ones [3]. In this context, genetic improvements of microalgal strains can play a major role in the development of biofuels technology. However, addressing such a complex problem via trial-and-error, as is done currently, presents many drawbacks. First of all, evaluating the performances of a mutant is a difficult and time demanding task, as it requires to grow several microalgae cultures under different light conditions and each experiment may last several days.

First principles models could be an effective tool to reduce the experimental effort needed to select the best mutant among different candidates and they can also provide guidance for genetic engineering by identifying those modifications having the largest potential impact on productivity.

In this work we will show how a model developed by the authors [4] can be exploited to evaluate the performances of different mutants in fast and reliable way. The model describes the photoproduction, photoregulation and photoinhibition processes and has been calibrated against chlorophyll fluorescence and photosynthesis rate measurements of three strains of *Nannochloropsis gaditana*: the wild type, a mutant with reduced NPQ (mutant T3.9) and a mutant with a reduced antenna size and chlorophyll content (mutant E2). After the calibration the model has been used to evaluate the mutants performances under multiple light conditions, such as the photoperiods of different months in a typical Mediterranean country. The advantage of the proposed approach is that to calibrate the model are necessary only fast and accurate fluorescence measurements and some measurements of photosynthesis rate. In fact, as demonstrated in [4] the information contained in fluorescence experiments is sufficient to describe the dynamics of photoregulation and photoinhibition.

In Figure 1a are reported the photosynthesis rate measurements for *N. gaditana* wild type along with the model predictions for the wild type, the mutant T3.9 and the mutant E2. In Figure 1b are reported the predicted photosynthesis rate profiles under light conditions typical of July in Padova as obtained from the PVGIS database.

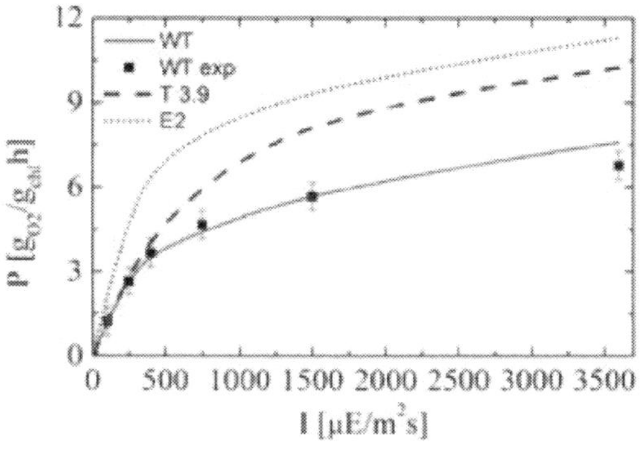

(a)

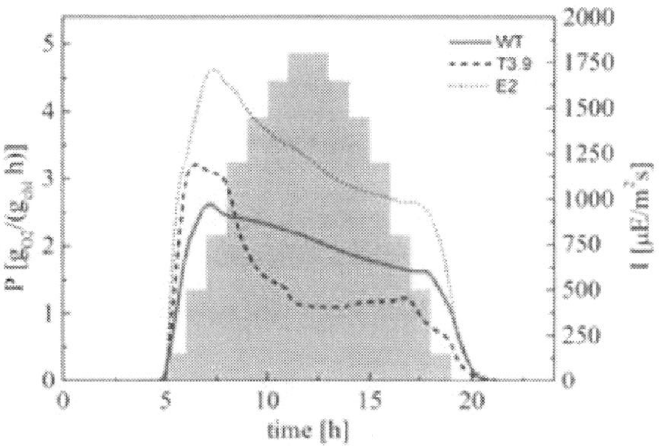

(b)

Figure 1: (a) Photosynthesis rate measurements for *Nannochloropsis gaditana* wild type along with model simulation results and photosynthesis rate prediction for mutants T3.9 and E2; (b) predicted photosynthesis rate profile for July irradiantion profile for *N. gaditana* wild type, mutant T 3.9 and mutant E2.

We can observe that, even if the photosynthesis rate measurements of the NPQ-less mutants are higher than the measurements of the wild type also at high light intensities, when the culture is exposed to intense light for a prolonged amount of time, the photoinhibition lead to a dramatic reduction of the photosynthesis rate. For this reason a mutant where the NPQ activity is absent may not be a good candidate for an outdoor cultivation in the Mediterranean area and other type of mutants should be investigated. On the other hand, the mutant E2 is more promising as the model predictions show an higher photosynthesis rate both in laboratory measurements conditions and for the July photoperiod. Moreover, if a dense culture is considered the antenna-less mutant will lead to a more uniform light distribution, which will increase the gap between the wild type and the mutant.

Future efforts will be devoted to validate the model predictions for different photoperiods and to include the light attenuation and mixing effect in the modelling framework. This will allow us to investigate the mutants performances under more realistic conditions representing a real photobioreactor conditions.

References

[1] Mata, T.; Martins, A.; Caetano, N. Microalgae for biodiesel production and other applications: a review. Renewable and Sustainable Energy Rev. 2010, 14, 217-232.

[2] Williams, P. J. l. B.; Laurens, L. M. L. (2010). Microalgae as biodiesel & biomass feedstocks: Review & analysis of the biochemistry, energetics & economics. Energy & Environmental Science, 3, 554-590.

[3] Formighieri, C., Franck, F., and Bassi, R. (2012) Regulation of the pigment optical density of an algal cell: filling the gap between photosynthetic productivity in the laboratory and in mass culture. J. Biotechnol. 162, 115–23

[4] Bernardi, A.; Nikolaou, A.; Meneghesso, A.; Morosinotto, T.; Chachuat, B.; Bezzo, F. (2015). High fidelity modelling of light limited photosynthetic production in microalgae using PAM fluorescence and optimal experiment design. PlosOne. 2016, 11, e0152387

Copyright © American Institute of Chemical Engineers. All rights reserved.

468558 Use of Unionized Ammonia to Control Zooplankton Grazers in Cultures of *Scenedesmus*

Abstract:

Algae have been targeted as a potential biomass source for scalable biofuel and bioproducts development. Unfortunately, the presence of zooplankton grazers has remained a problem in open pond systems, reducing yields and causing repeated culture crashes. At Old Dominion University's open pond, zooplankton grazers were controlled by maintaining a high pH (>9.5 pH) and total nitrogen concentration of 30 mg/L using urea. These values correspond to unionized ammonia values previously reported as toxic to the zooplankton, specifically rotifers, but not the algae in studies designed specifically for each organism. To understand this phenomenon further, particularly in the relevant combined (algae/zooplankton) system, 2.5 L cultures of zooplankton contaminated *Scenedesmus* were inoculated in bench scale photobioreactors at various pH's (8.0, 10.0 and 12.0) and unionized ammonia concentrations (0.0, 0.1, and 1.0 mM). These treatments were tested in a factorial experimental design. Trials lasting five days each were broken down by each ammonia treatment. Each trial was tested across each pH plus a control with only culturing media to ensure the algae cultures were growing normally (4 treatments per trial in duplicate). The algal and zooplankton populations were monitored by cell counts (hemocytometer) and total suspended solids (TSS) measurements (1.5 Âµm filter). Unionized ammonia was supplied in the form of ammonium chloride and the pH was daily adjusted using hydrochloric acid and sodium hydroxide additions. After the repeated crashes of the 12.0 pH cultures across all ammonia treatments, this pH treatment was removed from the experiment. To improve control over pH, phosphate and carbonate buffer systems were employed in the second set of trials, but the unbuffered systems were still tested for comparison. This resulted in 4 pH treatments plus a control in duplicate for each trial. Total ammonia was measured in filtered (1.5 Î¼m) liquid samples collected each day after TSS and frozen at -20 °C to quantify losses due to uptake by the algae and/or stripping to the atmosphere. Preliminary results indicate that 1.0 mM NH_3 concentration reduced and maintained a reduced zooplankton population for at least several days while slightly reducing the growth of the algae

population before ammonia concentrations were reduced to non-inhibitory levels.

Copyright © American Institute of Chemical Engineers. All rights reserved.

467182 Extracellular Glycerol Production ByÂ *Dunaliella Tertiolecta* in a Membrane Photobioreactor

Abstract:

Extracellular glycerol production by *Dunaliella tertiolecta* in a membrane photobioreactor

Photosynthetic microalgae can be used as bio-factories to produce value-added chemicals from carbon dioxide. One such chemical is glycerol, which is used as a precursor in many industrial manufacturing, including in the production of food and pharmaceuticals. Many marine microalgae, especially those belonging to *Dunaliella* sp., are able to produce intracellular glycerol under saline environment to regulate cellular osmotic pressure. Some microalgae, such as *Dunaliella Tertiolecta*, are also able to produce extracellular glycerol. While the use of microalgae in glycerol production has the advantages of carbon capture and utilization, it may also generate microalgae biomass, which can act as precursors in the production of various other useful compounds, including biofuels.

The extracellular glycerol production from microalgae is a great opportunity as it avoids the usual issues related to harvesting or downstream separation of microalgae. However, the production of extracellular glycerol by microalgae is typically low and it may not be very economical to produce. On approach to improve the yield of intracellular glycerol by microalgae can be the use of membrane photobioreactor (MPBR), such that a very high biomass density can be maintained in the MBR, which may translate in high yield of glycerol.

In this research, an MPBR based on microfiltration was designed and operated for high-throughput extracellular glycerol production using *Dunaliella Tertiolecta*. The effects of various operating parameters, such as biomass concentration, CO_2 supply, hydraulic retention time, light intensity and medium salinity on biomass accumulation and glycerol production were investigated. The results indicated that the MPBR has high potential in economical and sustainable production for glycerol.

Copyright © American Institute of Chemical Engineers. All rights reserved.

466442 Microalgal Fuels and Chemicals Production Using Kinetic Modeling and Scaled-up Experimental Studies

Abstract:

Microalgal Fuels and Chemicals Production using Kinetic Modeling and Scaled-up Experimental Studies Mesut Bekirogullari[a,b,c], Jon K. Pittman[c] and Constantinos Theodoropoulos[a,b,*]

[a]School of Chemical Engineering and Analytical Science, University of Manchester, M13 9PL, UK

[b]Centre for Process Integration, University of Manchester, M13 9PL, UK

[c]Faculty of Life Sciences, University of Manchester, M13 9PL, UK

Cultivation of microalgae for the production of biomass and lipid oils is a promising candidate in the field of renewable fuels [1,2]. The price of the microalgal biodiesel production is currently too high to compete with the fossil diesel production. Nevertheless, optimization and scaled-up control of the microalgae growth process can improve the competitiveness and sustainability of microalgal-derived biodiesel industry [3].

The aim of this work is the development of a detailed kinetic model which will be used in conjunction with in-house obtained experimental data to enable the optimization and the control of the microalgae growth in bench-scale batch systems. The proposed integrated framework will be employed to analyse the performance of microalgae growth, manifested by nutrient and environmental parameters such as substrate, nitrogen, light, temperature and pH. Successful control and optimization of such a system can lead to the utilisation of reduced excess/uncontrolled carbon, nutrient and water sources, and also to the production of high oil bodies.

A detailed ODE-based kinetic model is constructed here and it is used for the simultaneous simulation of the carbon substrate, the nitrogen, the light intensity and the pH in the system [4,5]. High fidelity experiments were conducted at a range of operating conditions to analyse the effect of substrate and nitrogen on the lipid oil production, and to complement the computational study in order to enable the estimation of system kinetic parameters of crucial importance. The constructed model is validated and it is utilised to define the optimal growth conditions, which lead to highest oil accumulation and biomass growth.

Furthermore, experiments were carried out in open raceway ponds under the obtained optimal growth conditions and they were used for kinetic model validation at a scale-up level. We believe that scaled-up process simulations using such a detailed modelling framework will enable the system industrial applicability and commercialization.

Keywords: microalgae, bench scale, raceway pond, multi-parameter quantification.

References

1. Driver, T., Bajhaiya, A. & Pitmann, J. K. 2014. Potential of Bioenergy Production from Microalgae. Current Sustainable/Renewable Energy Reports, 1, 94-103.
2. Y. Chisti, 2007; Biotechnology Advances, 25, 294-306.
3. Bechet, Q., Shilton, A. & Guieysse, B. 2013. Modeling the effects of light and temperature on algae growth: State of the art and critical assessment for productivity prediction during outdoor cultivation. Biotechnology Advances, 31, 1648-1663.
4. M. Bekirogullari, J. Pittman and C. Theodoropoulos, 2015; Computer Aided Chemical Engineering, 37, 2393-2398.
5. M. Bekirogullari, I.S. Fragkopoulos, J.K.Pittman and C. Theodoropoulos, 2016; Algal Research, *Submitted.*

* Corresponding author, *e-mail:* k.theodoropoulos@manchester.ac.uk

Tel: +44 (0)161 3064386, Fax: +44 (0)1613069321

Copyright © American Institute of Chemical Engineers. All rights reserved.

465404 Tertiary Wastewater Treatment By *Chlorella vulgaris* in a Membrane Photobioreactor: Effects of Wastewater Composition and Light/Dark Cycle

Abstract:

Microalgae is one of the most promising and sustainable feedstock for the production of biofuels because of its high growth rate, efficient CO_2 fixation and high lipid content. These advantages had led to extensive research in the microalgae cultivation and processing. However, despite making significant progress at these fronts, production of biodiesel from microalgae remains an expensive process.

One approach to reduce the cost of microalgae cultivation is the use of urban wastewater for microalgae growth. Municipal wastewater can not only provide cheap source of organic carbon to microalgae but it also contains ample amount of nitrogen (N) and phosphorus (P) as well. The cost of microalgal biofuels can also be reduced by achieving high biomass concentration in the photobioreactor which will result in lower harvesting costs. A high biomass concentration can be achieved in continuous cultivation of microalgae in a membrane photobioreactor (MPBR), by operating it at high loading rates.

In this research, an MPBR was designed and operated for N and P removal from wastewater using *Chlorella vulgaris*. The effects of N/P ratio and light/dark cycle on MPBR performance and biomass accumulation were investigated. The MPBR achieved nearly complete N removal under all the operating conditions, although the removal efficiency/rates were lower when the MPBR was operated under alternating light/dark cycles. It was also observed that P removal improved at higher N loadings. The biomass composition was different at different wastewater composition, and higher amount of proteins were accumulation by microalgae, when N supply was increased. The results suggested that autotrophic microalgae can be used for effectively treating and recycling inorganic pollutants in tertiary wastewater.

Copyright © American Institute of Chemical Engineers. All rights reserved.

452691 Modeling Algal Cultivation in the United Arab Emirates

Abstract:

Due to, first oil price volatility, and then climate change, Abu Dhabi has decided to become a worldwide leader in renewable sources of energy, sustainability, and carbon capture and sequestration (CCS). In 2006, Abu Dhabi launched the Masdar Initiative aimed at promoting the development and use of renewable sources of energy in Abu Dhabi [1]. This initiative includes the construction of Masdar City which will be a carbon-neutral city, the establishment of the Masdar Institute of Science and Technology, and the objective of increasing the share of renewables to at least 7% in Abu Dhabi's power generation by 2020 [1]. This has resulted in Abu Dhabi establishing a nuclear program for civilian use and the largest solar energy farm [2]. Abu Dhabi has also launched an ambitious CCS project that will have captured five million tons of CO_2 by 2012 [3], but there are concerns about the high cost of CCS. The main use of the sequestered carbon will be for Enhanced Oil Recovery (EOR), but Abu Dhabi is searching for other uses of this sequestered carbon. In addition, Abu Dhabi's city council data indicates that 82 million gallons of wastewater are dumped directly to the sea causing harmful algae blooms in the ocean.

These situations present an outstanding opportunity for algal cultivation processes because of their ability to fix atmospheric CO_2 and to remove nutrients from wastewater. Even more, biofuels and high value products can be obtained from algae reducing the cost of CCS. The United Arab Emirates (UAE) has a sizable potential for algal biofuels due to its vast desert expanses, long shorelines, year-round sunny days, availability of flue gas, seawater and wastewater, and native adapted algae species that tolerate salinity levels twice as high as those in the Gulf waters [2]. However, algal cultivation is challenging as it must be achieved in open ponds. Closed photobioreactors are not an option because they incur high capital and operating costs. Open pond systems have low biomass and lipids productivities because they are CO_2 limited and they are susceptible to invasion by undesirable microorganisms. In order to increase productivity of open pond systems, reliable and predictive process models are needed.

Here, we use the high-rate algal bacterial pond (HRAP) [4], [5] model and dynamic flux balance analysis (DFBA) [6], [7] to model algal cultivation in the UAE. The model and ideas presented in [8] are taken as a start point. In this paper microbial consortia are used for two purposes. First, they increase culture resilience by utilizing resources more efficiently compared to an algae monoculture, making pond invasion by undesirable species more difficult [9]. Second, symbiotic relationships can be established between microalgae and other microbes in which the microbes convert carbon sources into CO_2 and acetate, and algae converts the CO_2 and acetate into biomass. Therefore, algae ponds can become feasible in locations far away from flue gas sources.

Here, we present the results of different case studies concerning the cultivation of algae in Abu Dhabi. In particular, we use experimental data regarding environmental conditions, wastewater composition, and flue gas composition to test different cultivation configurations and make predictions of novel system behavior [10], [7], [11].

References

[1] D. Reiche, "Renewable Energy Policies in the Gulf countries: A case study of the carbon-neutral "Masdar City" in Abu Dhabi," *Energy Policy,* vol. 38, pp. 378-382, 2010.

[2] A. A. Jaradat, "The current potential of algae biofuels in the United Arab Emirates," *Biofuels,* vol. 4, no. 4, pp. 347-349, 2013.

[3] S. Nader, "Paths to a low-carbon economy - the Masdar example," *Energy Procedia,* vol. 1, pp. 3951-58, 2009.

[4] H. Buhr and S. Miller, "A Dynamic Model of the High-Rate Algal-Bacterial Wastewater Treatment Pond," *Water Res.,* vol. 17, pp. 29-37, 1983.

[5] A. Yang, "Modeling and Evaluation of CO2 Supply and Utilization in Algal Ponds," *Industrial & Engineering Chemistry Research,* vol. 50, pp. 11181-11192, 2011.

[6] A. Varma and B. Ø. Palsson, "Stoichiometric flux balance models quantitatively predict growth and metabolic by-product secretion in wild-type Escherichia coli W3110," *Applied and Environmental Microbiology,* vol. 60, no. 10, pp. 3724-3731, 1994.

[7] R. Mahadevan, J. Edwards and F. I. Doyle, "Dynamic flux balance analysis of diauxic growth in Escherichia coli.," *Biophysical Journal,* vol. 83, no. 3, pp. 1331-40, 2002.

[8] J. A. Gomez, K. Höffner and P. I. Barton, "From sugars to biodiesel using microalgae and yeast," *Green Chemistry,* vol. 18, no. 2, pp. 461-475, 2016.

[9] E. Kazamia, D. C. Aldridge and A. G. Smith, "Synthetic ecology - A way forward for sustainable algal biofuel production?," *Journal of Biotechnology,* vol. 162, pp. 163-169, 2012.

[10] J. D. Orth, I. Thiele and B. Ø. Palsson, "What is flux balance analysis?," *Nature Biotechnology,* vol. 28, pp. 245-248, 2010.

[11] K. Höffner and P. I. Barton, "Design of Microbial Consortia for Industrial Biotechnology," *Computer-Aided Chemical Engineering,* vol. 34, pp. 65-74, 2014.

Copyright © American Institute of Chemical Engineers. All rights reserved.

450165 Inhibitory Effect of Condensed and Non-Condensed Phenolic Moieties in Lignin on Enzymatic Hydrolysis

Abstract:

The residual lignin plays an inhibitory role in the enzymatic hydrolysis of cellulosic biomass. In this study, we examined the structure change of the isolated lignins from hot water hydrothermal pretreatment of aspen and their inhibitory effects on the enzymatic hydrolysis of Avicel. The functional groups of the isolated lignins were determined by quantitative ^{13}C, 2D HSQC and ^{31}P NMR. The increase of pretreatment temperature significantly increased the condensed and non-condensed syringyl and guaiacyl OH group contents in the isolated lignins, and decreased the aliphatic OH, p-hydroxybenzoate OH and carboxylic OH group contents. A compelling adverse association (r^2=0.998) was observed between the condensed syringyl and guaiacyl phenolic OH group contents in lignins and their inhibitory effects on enzymatic hydrolysis. Langmuir adsorption isotherms showed that higher pretreatment temperature resulted in higher binding ability between isolated lignins and enzymes, which led to more non-productive binding. It is hypothesized that condensed syringyl and guaiacyl phenolic units are mainly responsible for the inhibitory effect of lignin on enzymatic hydrolysis, in which the condensed aromatic rings enhance the hydrophobic interactions and the phenolic OH group boost the hydrogen bonding. The combination of hydrophobic interactions and hydrogen bonding can further strengthen the non-productive binding.

Copyright © American Institute of Chemical Engineers. All rights reserved.

467057 Measurement and Evaluation of Recalcitrance Changes for Organic Solvent Pretreatment Methods

Abstract:

Biomass recalcitrance is largely responsible for the high cost of chemical and fuel production from lignocellulose. To enhance biodegradation for these applications, it is crucial to determine how a pretreatment method affects the physical characteristics of lignocellulose[1,2]. In this work, we show the efficacy of Direct Yellow 11 as a probe to measure the cellulosic accessibility to cellulase enzymes. Together with crystallinity and delignification measurements, this accessibility measurement technique provides insight into recalcitrance changes that enhance biomass susceptibility towards enzymatic conversion to glucose[2,3]. Here we elucidate how a variety of organic solvents affect the key physical characteristics of biomass. These physical characteristic changes provide an increased understanding of biomass recalcitrance and its effect on cellulase kinetics.

1. Chandra, R., et al. (**2008**). "The characterization of pretreated lignocellulosic substrates prior to enzymatic hydrolysis, part 1: a modified Simons' staining technique." Biotechnology progress 24(5): 1178-1185.
2. Kang, Y. et al. (**2013**). "SO2 catalyzed steam explosion: The effects of different severity on digestibility, accessibility, and crystallinity of lignocellulosic biomass." Biotechnology progress 29.4: 909-916.
3. Hall, M. et al. (**2010**). "Cellulose crystallinity–a key predictor of the enzymatic hydrolysis rate." FEBS journal 277.6: 1571-1582.

Copyright © American Institute of Chemical Engineers. All rights reserved.

473939 One Step Breaking Softwood Recalcitrance By Cellulose Solvent-Based Lignocellulose Fractionation Followed By Enzymatic Saccharification

Abstract:

In our efforts to shift away from conventional petroleum-based feedstock to supply transportation fuels, plant biomass has emerged as an attractive renewable alternative. Softwoods have great potentials as renewable feedstocks for biorefineries. However, softwoods are more recalcitrant compared to hardwood counterparts. Hence, the use of softwoods in biorefineries has been viewed as limited. In the present study, we applied a cellulose solvent-based pretreatment/fractionation process on southern yellow pine and douglas fir at a mild condition: 50 °C and 1 atm for 90 min. Pretreated softwood samples were hydrolyzed fast by cellulase (15 FPUs) and > 85% glucose yield was achieved by 24 h. Cellulose accessibility to cellulase of pretreated samples was probed by the fusion protein and was found to increase by 39 folds. After fractionation, degree of crystallinity of pretreated samples decreased and fibrous structure appeared to be disrupted as observed by solid-state ^{13}C Cross-Polarization Magic Angle Spinning (CP/MAS) nuclear magnetic resonance (NMR) and scanning electron microscopy. No changes in lignin functionalities after fractionation were observed by 2D Heteronuclear Single-Quantum Correlation (HSQC) NMR, giving rise to the potential use of lignin as a co-product in biorefineries.

Copyright © American Institute of Chemical Engineers. All rights reserved.

472908 A Novel Acid Catalyst for Conversion of Ethanol to Butanol

Abstract:

The extensive consumption of fossil fuels contributes to a long-term threat to the ecosystem on which people highly depend. . Accordingly, people are increasingly interested in developing biorenewable fuels, among them is N-butanol. Due to its desired noncorrosivity, immiscibility with water, and relatively high energy density, preparing N-butanol has gained increasing attention. The major common challenge of the recently reported N-butanol production technologies lies in their low N-butanol selectivity. The novel acid catalyst (instead of base catalysts used in literature) developed in this work has achieved the highest selectivity to date. The effects of different factors on N-butanol selectivity and yield are evaluated. In addition, the associated reaction mechanism is understood with various material and process characterization methods.

Copyright © American Institute of Chemical Engineers. All rights reserved.

457412 Separation of Lignin from Organosolv Spent Liquor-from Phase Behavior to Continuous Processing

Abstract:

Worldwide pulping mills combust several hundred million tons of lignin every year in their recovery boilers to recycle the pulping chemicals and provide process heat [Cazacu et al. 2013]. The most applied Kraft and Sulfite pulping techniques are focused on pulp production whereas lignin is utilized as cheap energy provider.

Lignin is build up from phenylpropane monomers with hydroxyl and methoxy functional groups, making it a valuable renewable resource for the chemical and material industry. Lignocellulose biorefineries are supposed to sustainably provide renewable platform chemicals for the future by fractionating woody biomass with appropriate processes. Organosolv pulping techniques can fractionate lignocellulosic biomass into mainly pulp, lignin and (oligo-) sugars by using an organic solvent and water as pulping liquor at temperatures near 200 °C. The resulting pulp is of reasonable quality [Zhang et al. 2016] and the dissolved lignin can be separated from the spent liquor as precipitate by removing the organic solvent, which is recycled to pulping liquor. The (oligo-) sugar fraction remains dissolved in the aqueous phase. Organosolv lignin is known to be of high quality with low ash, sulfur and carbohydrate contents. The separation processes in a lignocellulose biorefinery utilizing organosolv pulping are essential for the economic viability [Viell et al. 2013].

The Fraunhofer Center for Chemical-Biotechnological Processes CBP in Leuna, Germany, operates a lignocellulose pilot plant consisting mainly of a batch forced circulation reactor for ethanol organosolv pulping, an enzymatic digestion unit for cellulose to sugar hydrolysis, as well as ethanol evaporation and lignin precipitation facilities. Fraunhofer CBP and MPI DCTS initiated a cooperation for investigation and optimization of the lignin separation process especially in terms of energy consumption, fouling prevention and downstream processability in filtration and drying.

In this contribution the path from fundamental phase behavior determination to an advanced continuous lignin precipitation process is presented [Schulze et al. 2016]. Systematic solubility measurements in the ternary lignin-ethanol-water system at different temperatures have been performed in order to elucidate attractive process possibilities. Thereby lignin was found to form a liquid phase under certain conditions as result of a decreased glass transition temperature. The liquid lignin phase potentially causes fouling on the reactor internals during separation processes if the process conditions are not carefully adjusted. On the other hand, softened lignin particles can agglomerate and coalesce to form bigger particles. A compromise between particle enlargement and incrustation formation has to be made in selection of process conditions. Enlarged lignin particles contribute to a better filterability. A (semi-) continuous process for simultaneous lignin precipitation, ethanol recovery and particle enlargement was developed on the knowledge from phase behavior. Experimental results from lab scale continuous processing of spent liquors produced at different pulping conditions indicate the influence of lignin properties (like molecular weight distribution) on the separation process. Semi-continuous precipitation experiments were performed in pilot scale confirming the scalability of the process to the next level. Inline process analytic techniques (Mettler Toledo FBRM®, PVM and ATR-FTIR) were utilized for monitoring and controlling lignin particle size distribution, lignin particle shape and aqueous phase composition. Agglomeration kinetics were derived from the monitored data. Filtration experiments with the produced lignin dispersions provided filter cake resistance data to assess filterability.

References:

Cazacu, G., Capraru, M., Popa, V. (2013). Advances concerning lignin utilization in new materials. In: Thomas, S., Visakh, P.M., Mathew, A.P. (Eds.), Advances in Natural Polymers. Springer, Berlin Heidelberg, pp. 255–312.

Schulze, P., Lorenz, H., Seidel-Morgenstern, A., Leschinsky, M., Unkelbach, G. (2014). Verfahren zur Fällung von Lignin aus Organosolv-Kochlaugen. DE priority patent application filed in 20th October 2014.

Schulze, P., Lorenz, H., Seidel-Morgenstern, A., Leschinsky, M., Unkelbach, G. (2016). Advanced process for precipitation of lignin from ethanol organosolv spent liquors. Bioresource Technology, 199, 128–134.

Viell, J., Harwardt, A., Seiler, J., Marquardt, W., (2013). Is biomass fractionation by Organosolv-like processes economically viable? A conceptual design study. Bioresource Technology, 150, 89–97.

Zhang, K., Pei, Z., Wang, D. (2016). Organic solvent pretreatment of lignocellulosic biomass for biofuels and biochemicals: a review. Bioresource Technology, 199, 21–33.

Copyright © American Institute of Chemical Engineers. All rights reserved.

462650 Cell Lysis and Lipid Recovery from Oleaginous Yeast

Abstract:

Single-cell oils, or lipids derived from oleaginous microbes, have seen recent interest as a potential renewable source of transportation fuels. However, one of the main technical challenges involved in single-cell-oil-derived products is due to the intracellular nature of the oils: the microbial cells must be lysed and the lipids extracted from the resulting complex matrix. We have screened several approaches for cell lysis approaches using the oleaginous yeast *Lipomyces starkeyi*, and have found that a dilute acid pretreatment compares favorably to a number of alternative strategies including high-pressure homogenization, alkaline pretreatment, bead beating, enzymatic hydrolysis or thermolytic approaches. Extraction of the lysed cells with hexane allows for recovery of the lipids in yields above 90% and in high purity. The acid pretreatment is also effective on strains of *Cryptococcus curvatus* and *Rhodosporidium toruloides*, and is highly scalable.

Copyright © American Institute of Chemical Engineers. All rights reserved.

471016 Transport and Reaction Modeling of Biomass Pretreatment and Topochemical Evolution Using Actual 3D Structure of Plant Cell Walls

Abstract:

Plant biomass is a renewable raw material for a wide variety of bio-based products which include pulp and paper, bio-plastics, biofuels, wood plastic composites etc. The process of conversion of plant biomass to other useful products is heavily dependent on our understanding of the physio-chemical structure of plant cell walls. Plant cell walls are composed of a complex network of pore spaces and fibers of cell wall components that has a significant influence on the efficiency of the biomass conversion process, the properties of the bioproducts and their end-use applications. Hence it is important to characterize the 3D plant cell wall structure. 3D characterization techniques such as X-ray computed tomography (XRCT) and Transmission electron microscopy computed tomography (TEM-CT) have been popular. Although, the X-ray CT provides micron scale visualization, it is not sufficient to get a complete understanding of the 3D structure of plant cell walls. Hence, a more powerful technique of TEM-CT is used to probe the nanoscale 3D ultrastructure. The computed tomography technique is based on capturing a series of images by successive tilts around a single axis and reconstructing the 3D image by back projection. The structural information obtained from TEM-CT can be correlated with topochemical information from Raman spectroscopy to fully understand the physiochemical structure of plant cell walls.

The first step in the conversion process is the pretreatment step, which disrupts the recalcitrant structure of lignocellulose and increases the efficiency of the subsequent hydrolysis. One or more of the cell wall components are dissolved during this process. The changes in structural characteristics such as porosity, specific surface area and 3D pore size distribution during the pretreatment process can be determined from the processed TEM-CT images. The structural information from TEM-CT can be correlated with the topochemical distribution from Raman Spectroscopy to develop a transport-reaction model based on a hybrid random walk and reaction. This model involves a stochastic dynamic approach that keeps track of the change in structure and spatial concentration of the cell wall components

simultaneously. A fixed number of particles of the reagent used for pretreatment, which is representative of its concentration, diffuse through the pore spaces of the plant cell walls till it encounters a pore-fiber interface. This path can be monitored by a stochastic hybrid random walk approach. At the interface, the reagent used for pretreatment can either react with one or more of the cell wall components or undergo reflection back into the pore space, based on the reaction probability. The, spatial concentration profile of the pretreating agent at different times can be determined by keeping track of the number of particles of the pretreating agent at each position as time proceeds. The spatial concentration profiles of the cell wall components and their evolution during biomass conversion can also be determined from the corresponding rate equations.

Thus, with such a model, a complete understanding of the 3D structure and the physicochemical distribution and its evolution can be understood. This provides a fundamental insight into biomass recalcitrance and how they can be addressed in developing efficient conversion strategies.

Copyright © American Institute of Chemical Engineers. All rights reserved.

472309 Lignin-Derived Products from Biocrude As Building Block Chemicals

Abstract:

The production of chemicals as bioproducts in an integrated biorefinery could enhance the economic viability of biomass-to-liquid fuel technologies and afford a higher return on investment. RTI International is developing an advanced biofuels technology that integrates a catalytic biomass pyrolysis step and a hydroprocessing step to produce infrastructure compatible biofuels. In recent efforts, RTI is exploring and developing processes that could be used to produce bioproducts in addition to liquid fuel from the biocrude intermediate.

The present study explores distillation as a separation method for recovery of lignin-derived chemicals from biocrude. The biocrudes used were relatively thermally stable and distillable; they were produced from catalytic pyrolysis with a non-zeolite, alumina-based catalyst. Gas chromatography-mass spectrometry (GC-MS) analysis indicated that the biocrude contained reasonable amount of lignin-derived compounds. The distillation studies were performed with a laboratory distillation unit (PETRODIST 300 CC) and the biocrude were fractionated into four distillate cuts: Cut 1 (IBP-110 °C), Cut 2 (110-200 °C), Cut 3 (200-300 ° C), and Cut 4 (300-400 °C). The yields, physicochemical properties, and chemical composition of the fractions as well as the overall distillation separation efficiency will be discussed.

Copyright © American Institute of Chemical Engineers. All rights reserved.

461248 Continuous Butanol Extraction Using Supercritical Carbon Dioxide

Abstract:

Butanol has promise as a drop-in biofuel or as a fuel additive that can be blended with gasoline at much higher proportions than ethanol. To date, economical production of butanol has been hampered largely by its cytotoxicity, which becomes limiting at levels as low as several weight percent. In this work, we demonstrate n-butanol extraction from dilute aqueous solutions used to simulate fermentation broths using supercritical carbon dioxide ($scCO_2$). In this talk, we will focus on process engineering of the n-butanol extraction. Our process engineering work is complemented by microbiological and metabolic engineering work that has identified a $scCO_2$-tolerant bacterial strain and genetically modified it for biofuel production. Compared to other butanol recovery approaches, $scCO_2$-extraction has sterilization and potential energy balance advantages; because $scCO_2$ selectively extracts butanol instead of water, a highly concentrated butanol stream can be recovered which requires minimal post-processing purification.

Specifically in this work, we studied n-butanol extraction performance and material/energy balance analysis. For butanol extraction performance, we used batch-wise extraction to study the effects of initial n-butanol concentration, extraction vessel pressure, and $scCO_2$ volumetric flow rate on n-butanol extraction rate. Additionally, we modeled the data using a standard liquid-liquid mass transfer model to determine values for the mass transfer coefficient, $k_l a$. Figure 1 shows representative extraction data compared to the model predictions using fit mass transfer coefficients. In all cases, the mass transfer model adequately described the experimental data. Best-fit values of $k_l a$ did not vary within our estimated limits of uncertainty for variation in extraction pressure (from 10.3 to 13.7 MPa – the range over which the $scCO_2$-tolerant bacterial strain has exhibited growth); therefore, we conclude that operation at lower pressures should be favored to achieve better process economy. Similar analysis was performed to interpret the mass transfer coefficient from correlations developed for gas-liquid and liquid-liquid extraction and compared to interfacial area results obtained from $scCO_2$-droplet size analysis. In addition to n-butanol, we investigated extraction of other alcohol products (e.g., iso-butanol, pentanol, and hexanol) as these are potential metabolic targets for a modified $scCO_2$-tolerant organism. Finally, we studied the extraction of the aldehyde intermediate, butanal, to set limits on its accumulation in the fermentation products.

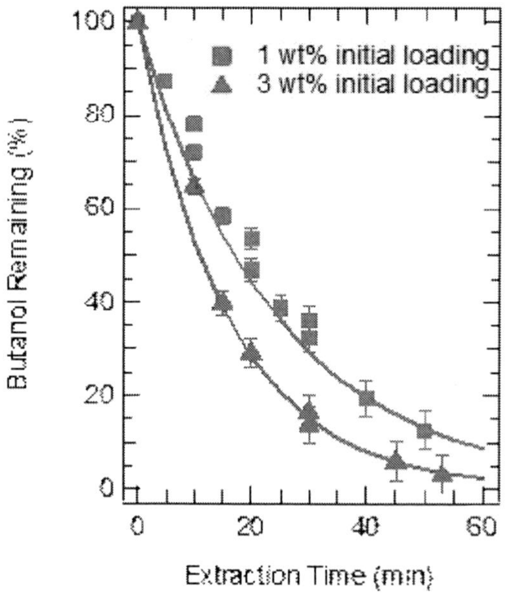

Figure 1. Representative n-butanol extraction data obtained at 10.3 MPa and 40 °C.

In addition to extraction experiments and modeling, we performed preliminary material and energy balance calculations of a conceptual continuous process. Thermodynamic and extraction data were used to model n-butanol removal from the fermenter and fermentation kinetics were modeled based on activity of similar organisms. The Peng-Robinson equation of state was used for modeling thermodynamic properties. To improve energy efficiency, the $scCO_2$ extraction stream was only partially de-pressurized from 10 to 5 MPa, resulting recovery of n-butanol at 79 wt% purity. Energy was recovered during the de-pressurization, which partly offset the energy required for CO_2 compression. Analysis was performed assuming the availability of either 10 or 0.1 MPa CO_2. Overall, we found that n-butanol with 95% purity could be produced at an energy cost of 3.9 MJ kg^{-1} (assuming that a 1 bar CO_2 source is available) or 3 MJ kg^{-1} (assuming a 10 MPa CO_2 source). These figures compare well with existing methods of butanol recovery, which range from 9 to more than 20 MJ kg^{-1}, and the lower heating value of the butanol fuel itself (33.1 MJ kg^{-1}). Further analysis is needed to account for heat loss and heat integration and to optimize process efficiency.

Copyright © American Institute of Chemical Engineers. All rights reserved.

466873 *Development of an Integrated Aquaculture System for Clean Water and Animal Feed Production*

Abstract:

The concept of a "simulated lichen system" was recently developed by our UMN research group in which we can select different desired microalgae and fungal combinations which can be grown on the surface of specific polymers to form a biofilm. This technology can be extended to aquaculture wastewater bioremediation in order to recycle the nutrients as a proteinaceous microbial biomass feed for the reared aquatic animals. This process modification could make aquafarming more efficient as commercial feeds account for up to 50% of total production costs. The "mycoalgae biofilm" utilizes photosynthetic aeration to replace the conventional and energy intensive mechanical aeration for biological processes. By using the mycoalgae biofilm in aquaculture production, farmers will realize lowered costs for feed and reduced costs for maintenance and cleaning of the aquaculture water. Microalgae cultivation has been heavily researched to treat nutrient-rich water; however, the process is limited by problems associated with the settling and separation of small microalgae cells downstream of the treatment site. The proposed methodology of using a bioaugment of filamentous fungi in the lichen biofilm overcomes this limitation. The fungi efficiently retain algae and can also aid in the recovery and recycling of nutrients in the wastewater. The fungi retaining the algae allows for a less energy intensive harvesting process as a singular solid biofilm is produced. The strains complement each other by the exchange of required respiratory gases. The biofilm composition can be tailored based on the influent stream components, feasibility of the strains to grow together and its nutritional value as animal feed.

The effect of different initial levels of ammonia were tested in synthetic aquaculture wastewater to study the removal efficiency of nutrient pollutants by our biofilm. Starting ammonia levels were chosen based on a literature review of effluent discharge from various cultured aquatic species. Ammonia levels were reduced to undetectable limits in the medium to low initial levels of ammonia trials. At high levels of ammonia a set ammonia removal rate was observed, independent of total ammonia present in the system. We suspect the removal rate is dependent on nutrient availability and can be optimized in the future. The mycoalgae biofilm outperformed mono-cultures of fungi and algae at every level of ammonia tested in regards to total nutrients removed from the media. Combined biomass production was also higher in the co-cultures than in mono-cultures at all ranges of ammonia tested. The biomass has added value as feed for the cultured fish, thus making the process more economically advantageous. The biomass generated will be high in polyunsaturated fatty acids, proteins, and carbohydrates thus enabling robust fish growth. The biofilm's nutritional properties, low energy harvesting method, and effective nutrient removal abilities make it a promising system for wastewater remediation.

Copyright © American Institute of Chemical Engineers. All rights reserved.

470075 Carbon Dioxide (CO_2) Capture and Utilization Pathways â€“ Assessing the Synergies Between Biological Use and Geological Sequestration of CO_2

Abstract:

As an important greenhouse gas (GHG), the emission of CO_2 to the atmosphere needs to be reduced in order to mitigate climate change. Designing of proper of CO_2 utilization pathways are imperative to reduce GHG emission. There are several ways to restrict CO_2 emission into the atmosphere. Two such promising technologies include geological sequestration of CO_2 in depleted oil and gas fields, saline aquifers and others, and biological sequestration of CO_2 via microalgae. The primary focus of the present study is to assess the integration potential of CO_2 geological sequestration and CO_2 biological sequestration via growth, harvest and conversion of microalgae to value-added bioproducts. Different ways of microalgae conversion, including lipid extraction, utilization of the spent algal biomass as well as hydrothermal liquefaction of the whole microalgal biomass, are considered for the production of value-added products. The present study aims to quantify the techno-economic tradeoffs and optimize the entire supply-chain of CO_2 utilization to ensure the most economically feasible alternative of CO_2 usage. A multi-scale supply chain optimization framework is considered which takes into account CO_2 capture, compression, pipeline transport and geologic storage as well as different processing pathways of microalgae. The state of Texas in USA is considered as the case study to evaluate the optimal design of CO_2 capture, storage and/or utilization pathways. The present study will help inform decision and policy makers the optimal pathway of CO_2 storage and utilization procedures for a particular location.

Copyright © American Institute of Chemical Engineers. All rights reserved.

AUTHOR INDEX

AUTHOR INDEX

AUTHOR INDEX